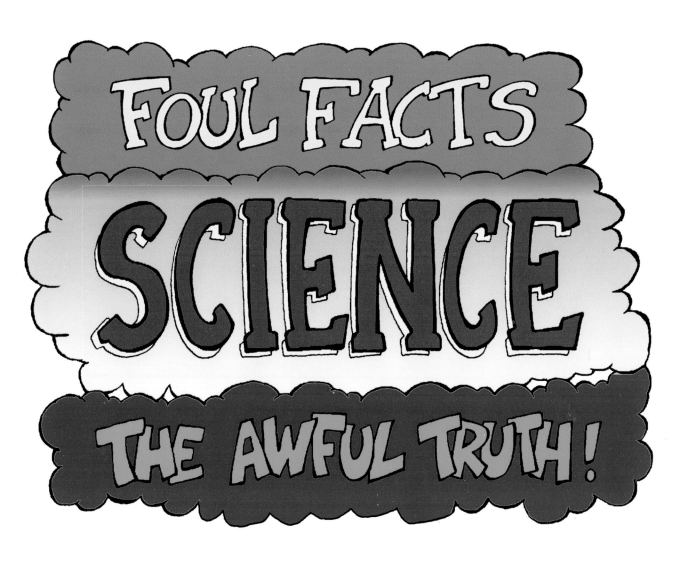

FOUL FACTS SCIENCE THE AWFUL TRUTH!

Martyn Hamer, Jamie Stokes and friends

Illustrations by

Mike Phillips

This is a Parragon Book
First published in 2001

Parragon
Queen Street House
4 Queen Street
Bath BA1 1HE, UK

Produced by Magpie Books, an imprint of
Constable & Robinson Ltd, London

ISBN 0-75255-302-X

A copy of the British Library Cataloguing-in-Publication Data
is available from the British Library

Printed and bound in China

CONTENTS

INTRODUCTION

Hello, my name is Albert. I love science, don't you? There are so many things that explode and make disgusting smells. I'm amazed they let youngsters like you do science in school at all, it's too much like fun.

This is a friend of mine, Professor Knowall, from the Institute of Really Smart Bods. He has agreed to show me around the Institute. Why don't you come along too? Apparently there is some pretty amazing and smelly stuff inside! Tell us Professor Knowall, what will we see?

First we will visit the World of Science Hall, where you can find out about a bridge named Galloping Gertie and exactly what happened to a woman who used Danforth's Non-Explosive Burning Fluid.

Next it's on to our Astronomy Department. Is it true that the Earth goes around the Sun? Find out who said so and why.

After that we'll be dropping in on the Medical Department. Meet Moldy Mary, discover why the Ancient Egyptians rubbed hippo fat on their heads, and learn about the funniest gas in the world.

Moving swiftly along we will come to our Nature Department. What exactly is the difference between a stalactite and a stalagmite, and why is it a bad idea to be a bald dinosaur?

Next it's time for some real fun as we stop a while in our Experiments Department. I'll be showing you some simple but exciting experiments that will let you make a real mess without getting into trouble.

After that it will definitely be time to visit our Mistakes Department where you will find out why making mistakes is sometimes the most useful thing a scientist can do!

To show you all just how useful science is we will then be strolling through our Inventions Department. Find out all about inventions that made it and a few that didn't! How about a luminous pocket sundial for use at night?!

Moving along we will come to the legendary Perpetual Motion Department. There are still lunatics in there who think they can make a machine run forever!

Since it's so much fun we'll backtrack to the Experiments Department and roll our sleeves up for some serious gunge and mess making. Have you ever made a gas gun or a balloon rocket? Come with us and I'll show you how.

Finally, after we've showered off and put the house back together we'll take a stroll through the Hall of Great Scientists. Meet the men and women who have changed our world.

A WORLD OF SCIENCE

He's Dam Mad!

Around AD 1000 the Egyptian scientist Alhazen came up with the idea of damming the upper section of the River Nile. This would prevent flooding in the lower reaches of the river and store water for irrigation. It would also have been useful to generate hydroelectric power but that hadn't been invented yet.

Caliph

The Caliph of Egypt thought the idea was brilliant and told Alhazen to go ahead. But when the scientist started to investigate the possibilities he found it just could not be done. He dared not tell the Caliph so he pretended to be mad – and he played the madman right up to the Caliph's death some twenty years later!

Science Can Be Deadly

The leaders of the French Revolution in 1789 were against the tax and banking systems devised by Antoine Lavoisier so they put him on trial and he was guillotined.

It was a sad day for science (and for Lavoisier of course) for Lavoisier was one of France's greatest scientists. He more or less founded the modern science of chemistry with several brilliant experiments in the 1770s. He was the first to show that air is mainly composed of two gases: oxygen and azote (later named hydrogen). In collaboration with other scientists he also devised the first logical system for naming chemical compounds.

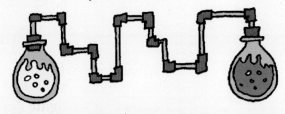

Man Ahead Of His Time

In addition to his importance as a scientist, Lavoisier took an active interest in public affairs, showing how improvements could be made in prisons, hospitals, agriculture and banking. He also helped found the metric system of weights and measures.

But perhaps he should have stuck just with science for the revolutionaries regarded him as an enemy of the people. "It took only a moment to cut off that head," said the mathematician Joseph Lagrange, "and perhaps a century will not suffice to produce another like it."

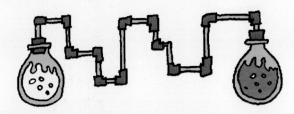

A Bad Case Of Wind

In 1973 a scientific research station at Cranfield, England had its roof blown off. The strange thing was that the establishment was set up to study the effects of the wind!

5

Journey To Disaster

When it was built in 1878 the Tay Bridge near Dundee, Scotland was the longest in the world. It was 3.2 kilometers (2 miles) long. Unfortunately the designer, Thomas Bouch, had got some of his figures wrong for he had underestimated the wind pressure blowing against the structure.

Windy Day

In a high wind in 1879 the bridge gave up the fight and collapsed just as a train was going over it. The train and seventy-five passengers plunged into the waters of the River Tay. Bouch never recovered from the shock and died the following year.

Galloping Gertie

The Tacoma Narrows Bridge at Puget Sound, Washington, USA, was designed to give in the wind. The design was so flexible, however, that it tended to sway even in light winds and locals nicknamed it "Galloping Gertie."

On 7 November 1940 Galloping Gertie galloped too much. It began waving like a snake and car drivers abandoned their vehicles and ran for their lives. Eventually the bridge was buckling so much in the wind it finally collapsed.

Scientists don't get it right every time!

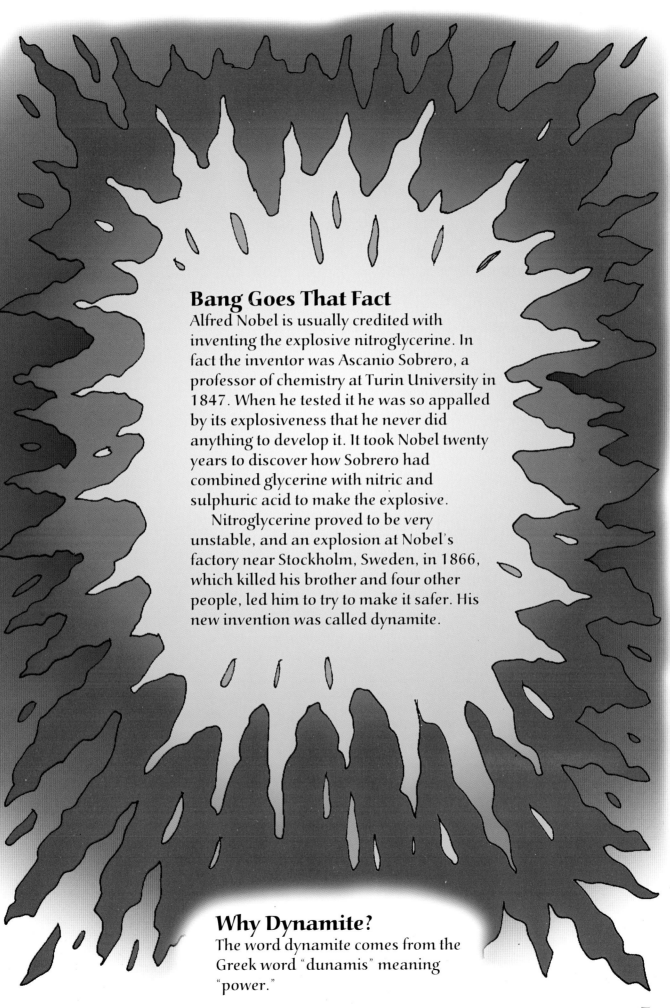

Bang Goes That Fact

Alfred Nobel is usually credited with inventing the explosive nitroglycerine. In fact the inventor was Ascanio Sobrero, a professor of chemistry at Turin University in 1847. When he tested it he was so appalled by its explosiveness that he never did anything to develop it. It took Nobel twenty years to discover how Sobrero had combined glycerine with nitric and sulphuric acid to make the explosive.

Nitroglycerine proved to be very unstable, and an explosion at Nobel's factory near Stockholm, Sweden, in 1866, which killed his brother and four other people, led him to try to make it safer. His new invention was called dynamite.

Why Dynamite?

The word dynamite comes from the Greek word "dunamis" meaning "power."

Shoddy Goods

Not all scientific products are as good as their makers claim. There is a gravestone in Girard, Pennsylvania, USA, that proves this. The epitaph to 26-year-old Ellen Shannon reads:

Who was fatally burned
March 21, 1870
by the explosion of a lamp
filled with "R.E. Danforth's
Non-Explosive
Burning Fluid"

Cold Chicken And Bacon

Science can be a dangerous occupation as the philosopher and essayist Francis Bacon discovered, to his cost, in 1626. One day he decided to stuff a chicken with snow (the chicken was dead, by the way) as an experiment in refrigeration. Unfortunately this simple experiment led to his death, as a short while later he developed bronchitis and died.

So beware the frozen chicken in your supermarket – cold chicken can kill! Maybe you should have a bacon sandwich instead?

Left Is Right

Most countries drive on the right-hand side of the road and it has been claimed by some scientists that this is what causes tornadoes. Perhaps the reason why there are so few tornadoes in Britain is that the British drive on the left?

Getting To The Seat Of The Problem

For many years wives have complained that their husbands do not put the toilet seat back down after they have been for a pee. Many brilliant minds have tried to find a solution to this age-old problem but without success.

In 1999 Terry Convoy, a surveyor from Harlow in England, announced that he had found the answer – a toilet seat that put itself down automatically!

The seat works by means of a small infrared sensor built into the underside of the seat. When the man has raised the seat and is doing what a man has to do the "magic eye" detects his presence and the seat stays in an upright position. When the man moves away the sensor releases a sprung hinge which moves the seat back down again.

It is early days yet but if the idea catches on automatic toilet seats could be found in every loo in the land.

SNAP! SNAP! SNAP!

Professor Knowall, why does the early history of science include so few women?

Up until the seventeenth century any woman who practised science took the risk of being branded a witch.

A Shocking Discovery

Luigi Galvani is credited with discovering electrical current in 1781, although many people believe it was his wife who first drew his attention to it. An electrical spark produced by a machine in Galvani's laboratory accidentally completed a circuit with a metal scalpel being used to dissect a frog, causing the frog's legs to twitch violently. This led Galvani to the discovery of what he called "animal electricity."

Later Alessandro Volta renamed it "current electricity." He was obviously an expert on current affairs.

Atomic BC

The atomic theory is believed to be quite a modern concept. Don't you believe it – the Greek philosopher Democritus of Adbera suggested that everything was made up of tiny invisible elements 2,400 years ago. But the idea is even older than that, for Democritus was simply developing ideas put forward by Leucippus of Miletus a hundred years earlier!

Mad Madge

Even in the seventeenth century, women who studied science were still considered rather strange. For example, Margaret Cavendish, the Duchess of Newcastle, expounded several theories about the nature of the universe. The reward for her work? She was nicknamed "Mad Madge of Newcastle."

A Matter Of Gravity

Anyone who saw Galileo Galilei climb to the top of the Leaning Tower of Pisa with two cannonballs in his hands in the year 1591 must have wondered what he was up to. And when he dropped the cannonballs from the top of the tower they probably thought he was mad.

But Galilei was not mad – he was testing a scientific theory. One cannonball was ten times heavier than the other but they both reached the ground at the same time. The fourth-century Greek philosopher Aristotle had taught that heavy objects fall faster than light ones but Galilei had proved him wrong. Galilei showed that the speed of a falling object depends upon the distance and not the weight.

On earth this experiment will not work with something extremely light and something extremely heavy because air resistance affects the rate of fall. But when the experiment was repeated on the Moon, where there is no air, a hammer and a feather fell together.

The Sceptical Chymist

The ancient Greeks and the early alchemists believed that earth, air, fire and water were the elements from which everything else was formed. In 1661 the Irish physicist and chemist Robert Boyle published a book called *The Sceptical Chymist*, in which he said that an element was a substance that could not be broken down into smaller parts.

It's A Gas

The word "gas" was invented by the Belgian chemist Jan Baptiste van Helmont, who lived from 1579 to 1644. He took it from the Greek word for "chaos" because the glass apparatus he used often broke during his experiments, a chaotic state of affairs.

Plaster Cast

No doubt you have come across plaster of Paris. It is the plaster used for setting broken limbs and making moulds. But did you know that it is really powdered gypsum or calcium sulphate, which is found in rocks, and that it gets its name from the gypsum quarries of Montmartre in Paris? When gypsum is heated to around 150 degrees Centigrade (300 degrees Fahrenheit), three quarters of its moisture is removed and it becomes powdery. When water is added it sets into a solid. So when you break your leg chemistry comes to the rescue!

Things Could Have Developed Faster

If the ancient scientists had not hung on to the idea that the Earth was the center of all things, the science of astronomy would have developed more quickly.

If medical men had not accepted the writings of the second-century AD Greek physician and philosopher Galen without question, medical science would have grown more rapidly.

If alchemists had not shrouded their experiments in secrecy, the science of chemistry would have developed earlier.

It just goes to show that scientists should always have an open mind and be prepared to question everything – even things that appear to be true.

13

The Search For Gold

Six thousand years ago humans knew how to obtain metals from the ground. Back then it was believed that it was possible to change one substance into another. The ideal metal was gold so the search began for a way of changing another metal, especially lead, into it. The men who attempted this feat were known as *alchemists* and their experiments, although mixed with a lot of mumbo-jumbo, formed the basis of what became the science of chemistry. Unfortunately, to keep their methods a secret they used strange symbols to represent the chemicals they used, so little is known of their discoveries.

Arabic Alchemy
The first alchemists appeared in Egypt and in fact the word "alchemy" comes from the Arabic "al kimia" which means "Egyptian art." From Egypt it spread into the Muslim world and then into Europe.

Many later alchemists claimed that they had managed to transform base metals into gold, but they were charlatans. If you can come up with a way of doing it you would soon become a millionaire.

Mystery Of The Emperor's Gold
When Rudolf II, Emperor of Germany, died in 1612 it was discovered that he had left his relatives four tons of gold. It was of the finest quality and cast into small ingots. The mystery was: where had it come from? Rudolf had been believed to be very poor but he was a keen supporter of the alchemists, so it was generally accepted that they had transmuted base metal into gold for him.

The Philosopher's Stone
According to the alchemists, the key to changing lead into gold was the Philosopher's Stone. This wasn't actually a stone but a powder, of which only a small quantity was required. It was believed to be made from mercury purified through hundreds of different processes.

A Strange Fact
One of the strange things about alchemy is that none of its most famous followers ever became rich men – so maybe there was nothing in it after all! But as all of the records kept by alchemists used a strange and secret language and symbols that no one can understand, perhaps we will never know.

Burning Books
In the sixteenth century the Swiss physician and alchemist Paracelsus burned books of alchemy in Basle, Switzerland, and announced that it was time to stop the futile search for gold and to use the methods of alchemy to aid mankind.

One of the first books on actual chemistry, as opposed to alchemy, was published by Andreas Libavius in 1597. But he did not get away from the old ideas altogether for his book was called Alchymia.

Name Change
The real name of Paracelsus, who lived from around 1493 to 1541, was Philippus Aureolus Theophrastus Bombast von Hohenheim. No wonder he changed it – but it would have been easier still if he'd just called himself Phil.

Momentous Moments

c 530 BC

Pythagoras works out his famous geometrical theorem. Schoolkids have been trying to figure it out ever since.

c 420 BC

Hippocrates establishes the Hippocratic Oath, which still forms the basis for the principles of modern medicine.

I CAN'T FIND ANYTHING IN THIS BOOK!

c AD 140

Galen publishes his works on anatomy, and it takes a very long time before people realize that a great deal of it is a load of old rubbish, largely because most of his studies were of animals, and not humans.

1600

William Gilbert publishes his works on magnetism because he thinks it is an attractive subject.

1614
John Napier publishes his tables of logarithms. Fred Bloggs makes his dining-room table from logs.

1781
The astronomer and telescope maker William Herschel discovers the planet Uranus. Born in Hanover in 1738, Herschel spent most of his life in England where he made numerous advances in our knowledge of the universe.

1796
Edward Jenner uses inoculation to fight smallpox. No one finds a cure for bigpox.

1899
Guglielmo Marconi transmits wireless signals across the English Channel. His friend Guglielmo Macaroni invents spaghetti.

1937
Frank Whittle designs the first jet aero engine then keeps whittling on about it.

ASTRONOMICAL ASTRONOMY

Astronomy is one of the oldest of the sciences. You have only to walk out of your cave at night to find out why. The heavens are fascinating, the stars magical, and the bright Moon must have been godlike – to say nothing of the Sun that appeared every morning (if it wasn't raining), moved magically across the sky, and then disappeared at the end of the day.

Hindu View

The ancient Hindus believed that the world was supported on the backs of four elephants, which were standing on a giant turtle. The turtle was in turn floating on a never-ending ocean. Just think what a shock modern astronauts would have had if this had turned out to be true!

Professor Knowall, do you know how the universe was formed?

It is generally believed that the birth of our universe came about some 15,000,000 to 25,000,000 years ago. At that time all the material and energy in the universe were concentrated in one massive globe. There was an enormous explosion, now known as "the big bang," which scattered matter in all directions until it eventually formed into stars. The galaxies are still moving away from one another and so the universe is still expanding as a result of that explosion.

Center Point

It used to be believed that the Earth was the center of the universe and that all the planets and stars revolved around it. It was obvious – all you had to do was look at the stars at night over even just a short period and you could see that they were moving around the Earth!

This was the general view until 1543, when a Polish astronomer named Nicolaus Copernicus upset the apple cart by suggesting that the Earth and the other planets went round the Sun!

Write It Down

Copernicus kept quiet about his ideas for some time because they were so revolutionary (literally) and he was scared of upsetting the Catholic Church. Eventually he was persuaded to put his ideas down on paper and his book *De Revolutionibus Orbium Coelestium (On the Revolution of Heavenly Spheres)* was published in 1543, the year that he died.

A Pretty Good Calculation

Way back in 1675 the Danish scientist Olaus Roemer calculated that light travelled at 186,000 miles per second. Almost 300 years later an American scientist made a more accurate measurement and found that light travelled at 299,792,458 meters (about 186,300 miles) per second. Considering how long ago Roemer did his calculations he was surprisingly accurate.

Proof Of The Pudding

It was not until the seventeenth century that there was any evidence that the Sun-centered system of Copernicus was correct. It was the Italian Galileo Galilei who, in 1610, first looked at the Moon and other heavenly bodies through a telescope and proved that Copernicus was right.

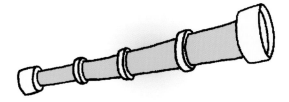

Arresting Theory

Galileo was immediately summoned to Rome because the Roman Catholic Church was adamant that the Earth was the center of all things. He was forced to say that the Church was right and that he and Copernicus were wrong – even though he was right, and he knew it! Then he was placed under house arrest for the rest of his life, his books were burned, and no one was allowed to consider his ideas.

It was to be another 360 years before the Catholic Church accepted the fact that the Earth does orbit the Sun.

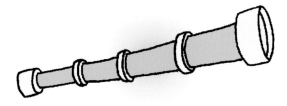

Homework Pays Off

Galileo Galilei did not have enough money to study mathematics at university so he studied at home. His hard work paid off for in 1589, at the age of just 25, he was appointed Professor of Mathematics at Pisa University.

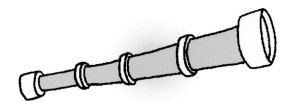

Scientific Smuggling

Even while under house arrest Galileo continued his work. His writings were smuggled out of Italy to the Netherlands, where they were published in 1638.

The Rough And The Smooth

Up until the time of Galileo everyone thought that the surface of the Moon was smooth. Galileo was the first person to observe that the surface was rough.

Let There Be Light

Distances in space are so great that measurement in everyday terms such as kilometers and miles gets rather cumbersome. Because of this astronomers measure distances in light years. A light year is the distance that light travels in one year. Light travels at 299,792,458 meters (about 186,300 miles) per second. Given that information you have no doubt already worked out that a light year is equal to 9,460,528,405,000 kilometers (nearly 6,000,000,000,000 miles).

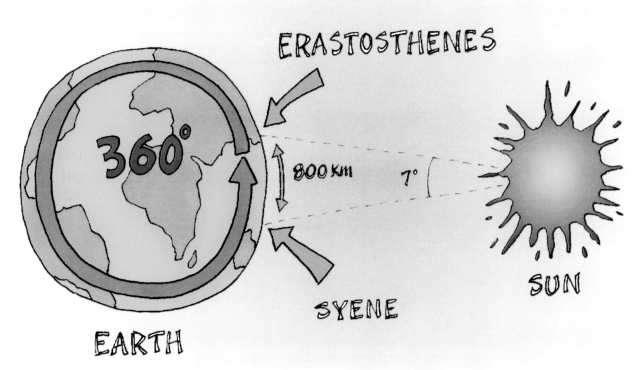

Going Around The World

Erastosthenes of Syene was a Greek astronomer and geographer in the third century BC. For many years he was in charge of a great library at Alexandria but he was also a bit of a brainbox. From the books he learned that the Sun was directly over the town of Syene (modern Aswan) on Midsummer Day. When he looked he found that it was not overhead at Alexandria but was seven degrees out. As a full circle contains 360 degrees, and seven is roughly (very roughly) one fiftieth of 360 he reckoned that the circumference of the Earth must be fifty times the distance between Syene and Alexandria. Erastosthenes found out that the distance was about 800 kilometers (500 miles) and from that calculated the Earth's circumference to be about 41,140 kilometers (25,700 miles). Quite amazingly he was only wrong by about 400 kilometers (250 miles), partly because he thought the Earth was a sphere. It's amazing what you can do with a bit of arithmetic and some thought.

Exactly As Predicted

Today Edmond Halley is best known for the comet named after him. But in his own time he became famous at the age of twenty-three when he published a book called Catalogus Stellarum Australium (Catalogue of the Southern Stars).

The top astronomer in Britain, Astronomer Royal John Flamsteed, encouraged Halley to continue his studies of the stars and planets. In 1682 a comet appeared and Halley predicted that it would reappear in 1758. Although he did not live to see it (he died in 1742) the comet did reappear in 1758 and is now known as Halley's Comet.

Halley's Comet

Halley's Comet had been seen several times before Edmond Halley was born but no one had realized that it was the same comet each time. Halley identified it as the comet seen in 1531 and 1607. Halley's Comet returns every 76 years so it will next be seen in 2062.

Comet In Cloth

The most famous appearance of Halley's Comet was in 1066. It was seen all over Europe and Matilda of Flanders included it in the Bayeux Tapestry (a pictorial account in needlework of the Norman conquest of England).

In The Round

Erastosthenes assumed that the Earth was a perfect sphere when he made his momentous calculations. But we now know that it measures more around the Equator than it does around the Poles. So it is a slightly flattened sphere. If you want to amaze your science teacher with your fantastic knowledge tell him (or her) that the Earth is an "oblate spheroid" – that's the official term for its shape.

Professor Knowall, what's the difference between a star and a planet?

Our Sun is a star and, like all other stars, it is made up of a mass of hot gases and radiates its own light. A planet, such as our Earth, moves around a star and is usually formed of a solid mass. A planet does not emit its own light but reflects the light from its star.

Sun Size

At one time in the dim and distant past it was believed that the Sun was only 70 centimeters (27 inches) in diameter. We now know it is a little bigger than that. Yes, you want to know how big it is – well, the diameter of the Sun at the equator is 1,392,000 kilometers (865,000 miles). OK?

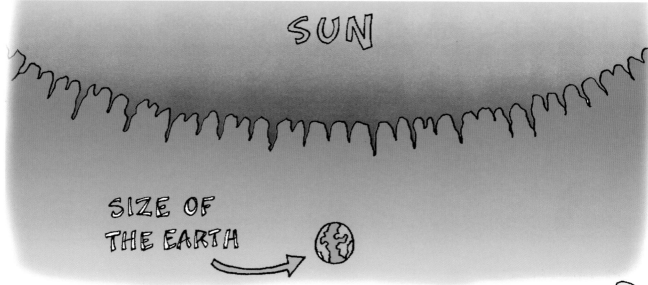

He Had A Nose For It

Tycho Brahe, who lived from 1546 to 1601, is regarded as one of the greatest astronomers before telescopes were used for heavenly observations. He made many incredibly accurate measurements of star positions and the movements of the planets at his observatory at Hven, an island in the Baltic Sea.

Although a brilliant astronomer Tycho Brahe was rather arrogant and perhaps a little tactless and he quarrelled with the Danish Court. Eventually he left Hven to become Imperial Mathematician to the Holy Roman Emperor, Rudolph II, in Prague, Czechoslovakia.

It was not the first time his attitude had caused problems and he had an unusual nose to prove it. As a student he had part of his nose sliced off in a duel so he made himself a false nose from gold, silver and wax!

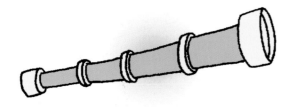

Before Copernicus

Copernicus was not the first person to suggest that the Sun was the center of our planetary system. In around 280 BC Aristarchus of Samos had put forward the same idea but very few people listened to his crazy suggestion. Perhaps he should have got his naked colleague Harry Starkers to streak through the streets proclaiming the news.

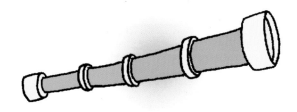

After Copernicus

The truth that the Sun is the center of our planetary system finally gained acceptance with the publication of *Laws of Planetary Motion* by the German mathematician Johann Kepler.

Kepler was lucky that he had access to the very accurate observations made between 1574 and 1596 by the great Danish astronomer Tycho Brahe and his sister, Sofie.

Getting The Needle

The hypodermic syringe was invented by Dr Alexander Wood of Edinburgh, Scotland. In 1853 he used the device to inject morphine into a patient. It was soon being used by many doctors and morphine given by injection was widely used as a treatment during the American Civil War. As a result 400,000 American soldiers became drug addicts, hooked on morphine.

Queen Shows The Way

The pain endured by women during childbirth was regarded as God's will until 1853. During the birth of her seventh child, Prince Leopold, on 7 April, Queen Victoria was administered chloroform by Dr John Snow. As a result other doctors began using analgesics during childbirth.

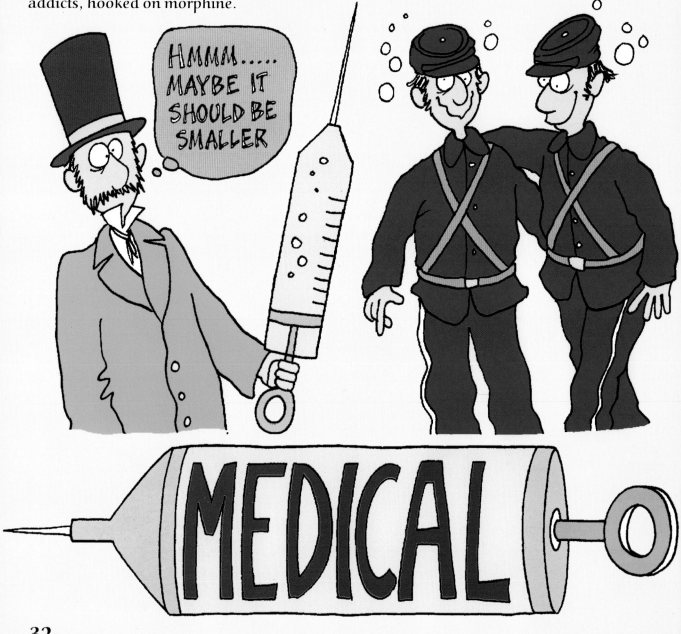

Urine Good Health

In days gone by many doctors made their diagnoses of a patient's illness by just looking at a sample of urine. They studied its colour, thickness, smell, and even taste (ugh!) and then told the patient what was wrong.

The technique was a load of old cobblers and had no scientific basis whatsoever. But don't say that to a modern doctor who asks for a urine sample. Modern doctors do not rely on guesswork any more: the urine is put through a series of chemical tests to give an indication of what could be wrong.

Mad Dog Disease

When a young boy called Joseph Meister was bitten by a mad dog with rabies he faced an inevitable and painful death. Rabies is a terrible disease that causes madness, convulsions and almost certain death. At the time the French scientist Louis Pasteur was working on a possible vaccine for the disease and he was persuaded to try it out on the young boy.

Joseph was given a dozen painful injections but he lived to tell the tale.

In The Ballance

From 1590 to 1620 an Italian professor called Santorio Santorio (perhaps his parents thought he was so good they named him twice) spent most of his time in a chair-like weighing device called a "Ballance." He wanted to find out how the body functioned so he weighed all the food he ate, all the air he breathed in, and everything that came out of his body – air, sweat, urine, and poo.

Fancy A Piece Of Sponge?

Unbeknown to his fellow parishioners the Italian scientist Lazzaro Spallanzani often went to church with bits of food under his armpits. These were not to provide him with a quick snack if he got hungry (you'd have to be pretty hungry to eat something covered in stinky armpit sweat), but to study how food is digested.

In other experiments on digestion Spallanzani would swallow small sponges tied on strings. After a while he would pull them out again to see what effect the stomach juices had on them.

Thanks To The Cow

Did you know that the medical word "vaccine" comes from a Latin word for "cow"?

It was fairly well known in the eighteenth century that people who worked with cows seldom caught the deadly disease of smallpox. Cows did get a milder form of the disease that was known as cowpox, and it was thought that people working with cows caught this disease and it somehow protected them from smallpox.

In 1796 an English country doctor, Edward Jenner, decided to test this theory. He took some fluid from a cowpox sore on the hand of dairymaid Sarah Nelmes and put it into the skin of a healthy eight-year-old boy called James Phipps. Six weeks later he deliberately infected the boy with smallpox. The boy did not get the deadly disease and Jenner was fêted as a brilliant medical pioneer. He must have breathed a sigh of relief, for if the boy had died he would now be famous as a sadistic murderer.

Thorn In The Side

Modern vaccinations are given with sterile syringes, but they had not been invented in Jenner's day so he used thorns!

Go To Hospital To Get Ill

In the middle of the nineteenth century it was strongly believed that something in the air caused wounds to fester after hospital operations. Almost half of the people who went in for an operation died from "hospital disease." Most people thought some unknown gas or invisible dust was responsible.

Joseph Lister, who was Professor of Surgery at the University of Glasgow in Scotland, discussed the problem with Thomas Anderson, Professor of Chemistry at the university. Anderson suggested that Lister read the findings of Louis Pasteur, who had just published his theory of disease from germs. Lister realized that these theories could apply to hospital operations so he set about thinking how he could kill the germs.

He could not use the heating method used by Pasteur so he had to find something else. Carbolic acid was the answer. It was used to dress wounds and also to clean the skin before operations. The success rate was so remarkable that it was not long before other surgeons were adopting Lister's ideas.

Lister Changed His Mind

Not everyone took to Lister's methods immediately. The Yorkshire-born surgeon John Wood said the new methods were the "delusions of a crank."

There was a heated argument between him and Lister in 1877 when a patient was brought in with a thigh tumor. The man's leg had to be amputated but Wood refused on the grounds that "poisoning will occur within five days and he will be dead."

Lister performed the amputation, much to Wood's disdain. Five days later Wood was passing the ward as Lister was removing the man's dressings. There was absolutely no sign of poisoning. Wood went over to Lister and shook his hand, "Tomorrow," he said, "I want you to teach me your methods." Lister did and the men became good friends.

EXCUSE ME, DO I GET A SAY IN THIS?

Thanks To Moldy Mary

Moldy Mary was the nickname given to a lady whose job in 1943 was to go round the markets buying mouldy melons. Sounds like a weird job, but Mary worked for the Northern Regional Research Laboratory at Peoria, Illinois, USA. The bad fruit was needed because it produced a mold called *penicillium chrysogenum* from which the drug penicillin could be cultivated. The laboratory was working with the scientist Alexander Fleming, based in England, who had discovered the medicinal use of penicillin. He had found that the mold from the fruit was cheaper and easier to produce than the process he had been using.

After Fleming

Strangely enough, although he realized it could be used to save lives, Fleming never used penicillin on humans. That was left to Howard Florey and Ernst Chain some ten years later.

By 1943, when war was raging in Europe, penicillin was being developed on a large scale and used to treat wounded soldiers. It saved many lives and is still one of the major drugs used today in the ever-continuing battle against disease.

Even Scientists Make Mistakes

In 1871 Joseph Lister was looking for something that would kill bacteria. As part of his experiments he tried to grow the bacteria that caused typhus and cholera but came up with something different. The mold he developed was called penicillium so he experimented with that instead. He found that germs grown on a thin layer of penicillium grew quite normally, but if they were cultivated on a thicker layer they did not reproduce. Unfortunately Lister did not realize the importance of his findings and went on to something else. If he had persevered he could today have been acclaimed as the discoverer of penicillin.

Discovered And Discarded

Ernest Duchesne could have been famous but for one small mistake. He was a teacher at the French Army Medical Academy at the end of the nineteenth century and he experimented with the mould penicillium. He noticed that when he put bacteria on the mold they died very quickly. In another experiment he injected laboratory mice with bacteria. Half of the mice were injected with penicillium. The treated mice survived but the others died.

"Interesting," thought Duchesne, but then did nothing more about it. If he had just gone on a little further he would have been famous today as the discoverer of penicillin.

Davy Was A Laugh

If you wanted a good laugh in the early 1800s you could go along to one of the gatherings organized by the British chemist Humphry Davy. In 1799 "Hilarious Humph" was combining gases to produce nitrous oxide. He accidentally inhaled some of the mixture, an accident that made him quite happy. It was not long before people were flocking to try some of this laughing gas for themselves.

It was during these hilarious happenings that Davy discovered that people who had inhaled the gas did not feel pain. He tried to persuade doctors to use it but had little success, and it was to be another forty years before anesthetics of various types became generally accepted as a way of alleviating pain during surgical operations.

Bread Cure

In ancient Egypt a cure for a scabby head was to rub it with hippopotamus fat and moldy bread. Strangely enough it worked. What the Egyptian doctors did not know was that they had discovered penicillin. If they had realized this Alexander Fleming would not be so famous today.

Listen To The Beat

A young French doctor called René Laënnec got the idea for the stethoscope from watching some children at play. Some were making scraping noises at one end of a fallen tree trunk and others were listening at the other end.

Laënnec thought he could use this idea in his work so he rolled a sheet of paper into a tube and held it against a patient's chest. When he listened at the other end of the tube he could hear the patient's heart beat! Later he made wooden tubes – the first stethoscopes.

Make Your Own

A simple stethoscope can be made from a length of rubber tubing and a small funnel (not a ship's funnel but a conical device for pouring liquids). If you can't find a funnel you can make one by cutting the top off a washing-up liquid bottle.

Push the stem of the funnel into one end of the rubber tube. Place the funnel end of your stethoscope against someone's heart and hold the other end to your ear. You should be able to hear the heart beating. You are now well on the way to becoming a doctor!

Exploding Trousers

In a cartoon published in 1802 James Gillray showed one of Humphry Davy's demonstrations. It depicts Sir John Coxe Hippisley, manager of the Royal Institution, sampling the nitrous oxide. It must have been a very effective demonstration, for the picture has flames and smoke bursting from the rear of Hippisley's trousers!

Give It Some Stick

Another experiment you can try is similar to what René Laënnec probably did after he had seen the children playing. Get a long stick and place one end of it to touch a clock or anything else that makes a fairly low noise. Hold the other end of the stick against your ear. You will be able to hear the clock much louder than normal because the wood carries sound waves much better than air.

Be careful not to poke the stick in your ear or you will end up hearing nothing at all.

If It Smells Strong It Must Be Good

Carbolic acid was originally used as an antiseptic purely because of its strong smell. Rotting wounds smelled and foul drains smelled, so it was thought that the strong-smelling carbolic acid must be the answer.

The sewers had been blamed for outbreaks of diphtheria and typhoid, and for cattle getting sick in fields where sewage was dumped. Carbolic acid was poured into the sewers and it got rid of the problems although no one knew why.

Willow Cure

When aspirin was first developed by the German chemist Felix Hoffmann in 1897 the salicylic acid used to make it came from the bark of the willow tree.

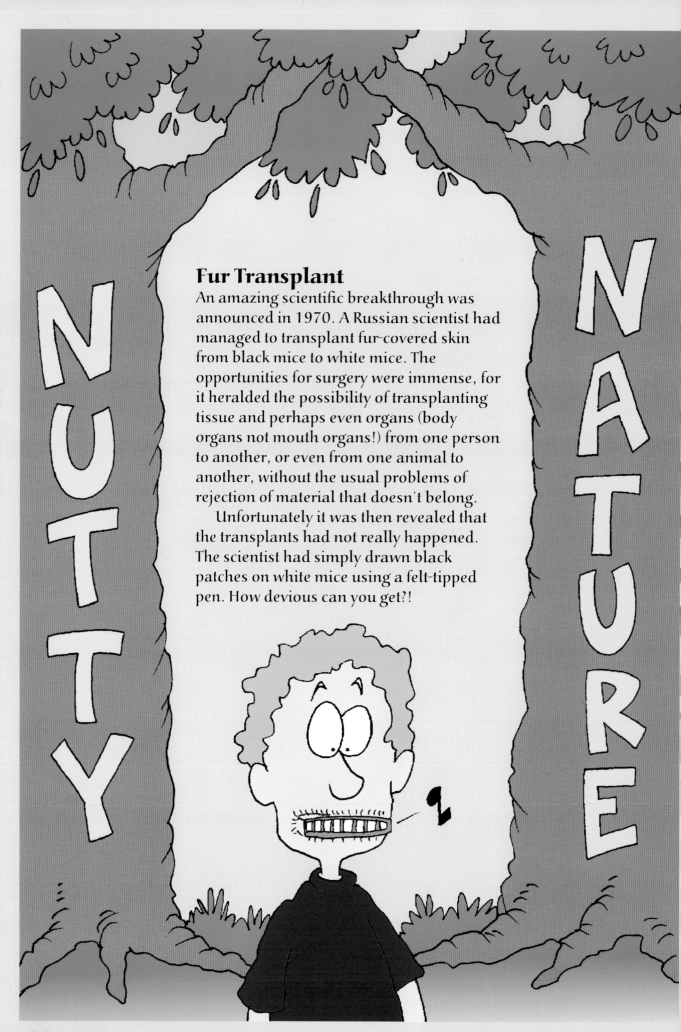

Fur Transplant

An amazing scientific breakthrough was announced in 1970. A Russian scientist had managed to transplant fur-covered skin from black mice to white mice. The opportunities for surgery were immense, for it heralded the possibility of transplanting tissue and perhaps even organs (body organs not mouth organs!) from one person to another, or even from one animal to another, without the usual problems of rejection of material that doesn't belong.

Unfortunately it was then revealed that the transplants had not really happened. The scientist had simply drawn black patches on white mice using a felt-tipped pen. How devious can you get?!

Two-Tone Flower

Plants take up water and nutrients through thin channels in their stems. To do this they have to defy the law of gravity and make everything flow upwards. This is done by a process called *capillary action*.

To show capillary action at work take two containers of water. Color one of them with blue ink and the other with red ink. Now take a white rose, dahlia, or carnation and carefully split it down the stem. Place one part of the stem in the blue water and the other in the red.

In just a few hours half the flower will be blue and the other half red as the colors have been sucked up through the stems.

And if you read the title of this experiment as "two-ton flower" and were expecting something weighty – tough!

Professor Knowall, why is the sea salty?

The salts in the ocean are the result of over two billion years of disintegration of rocks by rivers and rainwater. Soluble materials, such as salt, remain in the water but the insoluble materials form ocean sediments and rocks. One liter (about 1.8 pints) of sea water contains about 25 grams (nearly an ounce) of salts, of which some 18 grams (0.6 ounces) are common salt (sodium chloride).

Coming Up And Going Down

If you have ever been underground or in a large cave you will probably have seen stalagmites and stalactites. If you haven't seen any do not despair – you can make your own! And, what is more, you do not have to venture underground to do it.

All you need are two tumblers, a piece of cotton, some water, a plate, and some Epsom Salts.

Fill the tumblers with water and dissolve as much Epsom Salts in it as possible. Now put one end of the cotton in one tumbler of water and the other end in the other tumbler. Place the plate between the two tumblers, at the lowest point of the cotton.

In reasonably warm conditions the water will travel along the thread by capillary action. At the middle point of the cotton the water will drip down onto the plate but some of the salt crystals will remain on the thread to form a miniature stalactite, and where the water drips there will be a build-up of salt to form your stalagmite.

Stalagmites and stalactites are formed in the same way (although there are no tumblers and thread involved). Water dripping from the roof of a cave contains lime or other minerals. As the water evaporates some of this matter remains, either hanging from the roof or deposited on the ground.

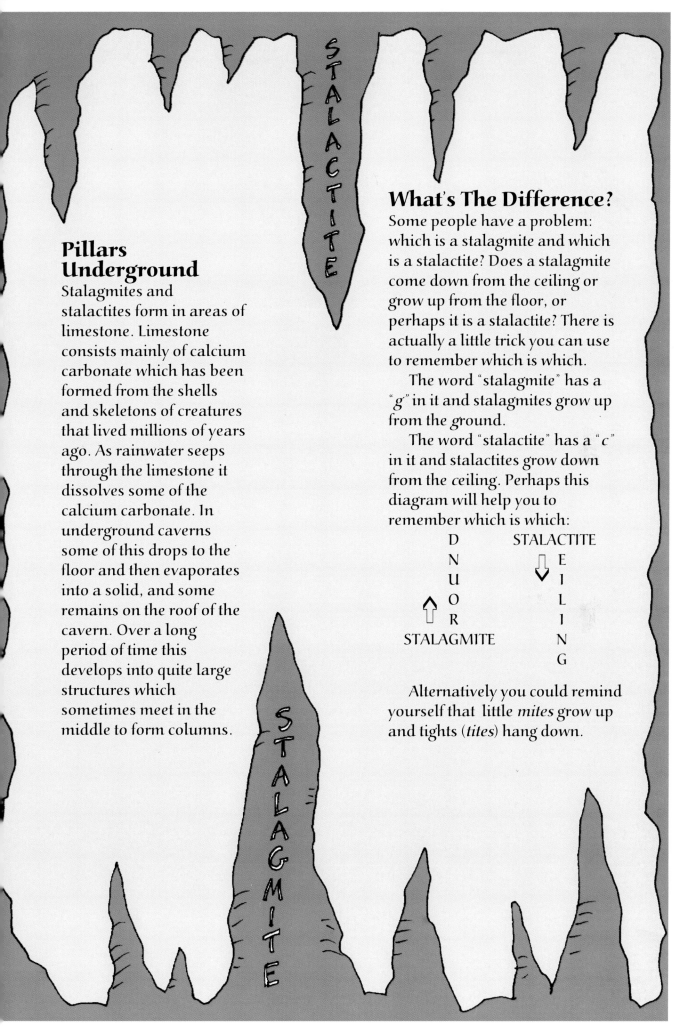

Pillars Underground

Stalagmites and stalactites form in areas of limestone. Limestone consists mainly of calcium carbonate which has been formed from the shells and skeletons of creatures that lived millions of years ago. As rainwater seeps through the limestone it dissolves some of the calcium carbonate. In underground caverns some of this drops to the floor and then evaporates into a solid, and some remains on the roof of the cavern. Over a long period of time this develops into quite large structures which sometimes meet in the middle to form columns.

What's The Difference?

Some people have a problem: which is a stalagmite and which is a stalactite? Does a stalagmite come down from the ceiling or grow up from the floor, or perhaps it is a stalactite? There is actually a little trick you can use to remember which is which.

The word "stalagmite" has a "g" in it and stalagmites grow up from the ground.

The word "stalactite" has a "c" in it and stalactites grow down from the ceiling. Perhaps this diagram will help you to remember which is which:

```
        D              STALACTITE
        N
        U                    ⇓  E
        O                       I
        R                    ⇑  L
                                I
STALAGMITE                      N
                                G
```

Alternatively you could remind yourself that little *mites* grow up and tights (*tites*) hang down.

The Missing Link

It has long been accepted that humans are descended from apes. In the early 1900s there was a lot of scientific discussion (that's hot air to us lesser mortals) about the *missing link*, a creature that was part ape and part man. The proof came in 1912 when an amateur fossil hunter, Charles Dawson, found some interesting bones in Piltdown, Sussex, England.

Among the bones were parts of a skull which obviously housed a large, human-like brain, and yet the jaws and teeth were those of an ape. Here at long last was the missing link that scientists had searched for.

There was some doubt about the finds at the time but it was not until 1953 that chemical dating proved that Piltdown Man, as the owner of the skull had been named, was a fake. The skull bones were human but only a few hundred years old, and the jaw and teeth were from an orang-outang and had been treated to make them look old.

Just proves that even scientists can be fooled.

First To See Bacteria

In 1674 the Dutchman Anton van Leeuwenhoek looked at a drop of canal water through a microscope he had made. Much to his surprise he found it teeming with tiny creatures. He thus became the first person to see bacteria, which he called "animalcules."

Getting To The Top

In spite of the fact that he knew no English and had no scientific education Anton van Leeuwenhoek was elected a member of the Royal Society of London in 1680. The Royal Society, the most famous scientific society of its day, was incorporated by King Charles II in 1662.

Do It Yourself

The microscopes used by Anton van Leeuwenhoek were made by himself and were nothing like the microscopes used today. His consisted of a small bead ground into a lens, which was fixed into a hole in a brass plate. He made over 400 microscopes, the most powerful of which could magnify objects 275 times.

A Better View

The British scientist Robert Hooke invented the compound microscope, the type of microscope with which most of us are familiar. This had two lenses in each end of a tube and produced a much better view than the single lens microscope used by van Leeuwenhoek.

The Dinosaur That Wasn't

Scientists all over the world were over the moon about the dinosaur discoveries of Dr Albert Koch in 1895. Koch had found in Missouri, USA, the bones of a marine python which had lived some twenty million years ago. He had found so many bones from the creature that an American museum had advanced him enough money to have it assembled into a complete skeleton.

The 100-meter-long (328 feet) skeleton, now named *python Missourium*, was exhibited all over Europe – until it reached the German city of Dresden. There a professor of zoology asked if he could examine the bones. When he did he discovered that they were nothing more than a collection of bones from several different animals and Koch was forced to admit that the whole thing was a fake.

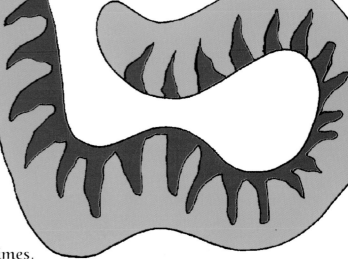

Germs As Big As Birds

Gustavas Katterfelto, an eighteenth-century entertainer who mixed conjuring, science and fraud in equal quantities, boasted that he had a "solar microscope" that could enlarge germs until they were as big as birds.

Katterfelto claimed he had invented the solar microscope – what he had was in fact a type of projector, using sunlight as its light source, invented by Johannes Liberkühn, which showed living organisms, such as maggots and small insects, on a screen.

Fossils Frowned On Fossils

To the old fossils of the scientific world of the seventeenth century Robert Hooke's views about fossils were revolutionary. It had always been believed that fossils were formed naturally, like crystals, or had been put on Earth by God. Hooke suggested that they could be the remains of living things. Shock! Horror!

Even scientists, who should be open-minded about all things, can be "blind" at times.

Animal Matters

In the eighteenth century the Swedish botanist Carl von Linné had problems because previous botanists had made mistakes in their descriptions of plants. He began devising a system that would classify all plants in an orderly fashion. He established what is now known as the binomial (two-name) system and published his findings in his book *Species Plantarum* in 1753. He then turned his attention to the animal kingdom and his system of animal classification was described in his *Systema Naturae*, published in 1758.

Although different methods are used today the systems outlined by Linné provided the basis for all future work.

Even The Botanist Had Two Names

Because he wrote his books in Latin, Carl von Linné is also known by the Latin name of *Carolus Linnaeus*.

His parents actually wanted him to be a clergyman and one of the reasons he did not take up this profession was because he hated learning Latin!

Linnaeus' System

In Linnaeus' system all living things are placed into seven main groups called taxa. These groups are:

KINGDOM

This is the largest division of all living things. There were just two kingdoms – animal and plants, but scientists later added a third, *Protista*, to cover living things that are neither plant nor animal but which have some characteristics of both.

PHYLUM

A group of related classes.

CLASS

A group of related orders.

ORDER

A group of related families.

FAMILY

A group of allied genera.

GENUS

A group of closely related species, although some genera (the plural of genus) consist of just one species.

SPECIES

A group of living things having common characteristics and which are capable of cross-breeding.

What Is Man?

Under the classification system humans are known as *homo sapiens*.

This comes about as follows:

KINGDOM

animal

PHYLUM

chordata (animals with a backbone)

CLASS

mammalia (mammals are warm-blooded animals that suckle their young)

ORDER

primate (the highest order of animals)

FAMILY

hominidae (the family of man)

GENUS

homo (human)

SPECIES

sapiens (the wise)

A Bit Of A Mouthful

Using every group to describe an organism would get rather chaotic so scientists usually just use the genus and species. That is why the official term for man is *homo sapiens*.

B-lime-y!

In London, England, there is a Linnean Society of biologists who preserve the work of Linnaeus and study developments in the science. Its crest incorporates a lime tree because Linnaeus' Swedish name, Linné, means "lime tree."

Savior Of The Silk Worm Industry

In 1865 the French silkworm industry was in trouble because a disease was affecting the silkworms. The scientist Louis Pasteur came to the rescue by finding out what kind of germs were causing the disease and how they could be eliminated.

ANIMAL - CHORDATA - MAMMALIA - PRIMATE - HOMINIDAE - HOMO - SAPIENS OR REGULAR GUYS AND GIRLS

Whining Winemakers

The winemakers of France were whining a great deal in 1856 because their wine was going sour and they didn't know why. They called Louis Pasteur for help. Pasteur examined some of the wine under a microscope and discovered various yeast microorganisms in it. Some had to be there to help in the fermentation process, but others were causing havoc and souring the wine. Pasteur advised the winemakers to heat the wine to kill these organisms. They followed his advice and instead of whining and crying they ended up happily wining and dining.

Have A Glass Of Hot Milk

In 1862 Louis Pasteur found that when milk was heated to just 70 degrees Centigrade (158 degrees Fahrenheit) the bacteria in it were killed, the milk would keep for several days, and the flavour was unaffected.

All Greek To Owen

The word "dinosaur" comes from two Greek words: "deinos," meaning "terrible" and "saurus," meaning "lizard." It was devised in 1842 by Richard Owen, the first director of the Natural History Museum in London, because of all the bones that were being discovered in England at that time.

First Dinosaur Discovery

The seventeenth-century professor of chemistry, Reverend Robert Plot, was the first person to record the finding of a dinosaur bone. In his *Natural History of Oxfordshire*, published in 1677, he described a large thigh bone found in a quarry. But of course no one had ever heard of dinosaurs at that time so Plot thought it was from a giant human being. Oops!

Marvellous Meglasaurus

The bone described by Robert Plot was actually the thigh bone of *Meglasaurus*, which was the first dinosaur to be given a name. It was named by the English geologist William Buckland in 1824.

Although a clever man and a devoted churchman (eventually becoming Dean of Westminster), Buckland had one strange habit. He had a pet bear that he would take to garden parties, the animal being correctly dressed in a cap and gown as Buckland was a geologist at Oxford University.

Baldness Beat Them

One theory as to why dinosaurs died out is that they were bald – they had no body hair – so, as the climate of Earth got colder, they were unable to keep warm enough to survive.

Terrible Tyrannosaurus

We tend to think of dinosaurs as enormous and rather stupid creatures but in fact some dinosaurs were quite small, no bigger than a chicken. One that *was* big was *Tyrannosaurus*, possibly the largest flesh eating animal ever to walk the Earth. Its weight was equivalent to two elephants and when raised up to its full height it was around 5 meters (16 feet) tall. Its head alone was over 1¼ meters (4 feet) long and its jaws were jam packed with sixty ferocious teeth some 15 centimeters (6 inches) long, with sharp serrated edges – they could certainly give you a nasty bite!

Voyage Of The "Beagle"

In 1831 HMS *Beagle* left England to survey the coast of South America. The biologist on board was 22-year-old Charles Darwin, who was later to shock the world with his theory of *evolution*, in which the process he called "natural selection" determined how all living things developed. He said that the development of a species was a slow, continual process, and that humans had evolved from ape-like creatures. These theories caused a great deal of fuss when they were published, especially among religious people who believed literally in the biblical story of creation, in which humans were put on Earth in their present form.

Meal In A Dinosaur

On 31 December 1853 twenty-one scientists had a meal in a dinosaur – or at least a model of one. It was a model of an iguanodon made for the Crystal Palace exhibition in London.

How Does A Scientist Work?

Scientists are trying to find answers to things that puzzle them.

They often do this by first observing whatever is causing the puzzlement. This may take some time because the problem has to be looked at in as many different ways as possible.

Next the observations are recorded. This can be in the form of written reports, photographs, videos, X-rays, mathematical tables and so on.

Then all the available information is studied and a possible answer to the problem is thought up. This possible answer is called a *theory*.

The next stage is to try out the theory to see if it is right. If it is not, then it is usually back to square one to start the whole process over again.

This procedure is called the "scientific method."

Here are some experiments you might like to try, to practise your powers of observation.

Color Magic

Get an adult to help you with this experiment as it involves bleach, which can burn your skin or damage your clothes. Pour some water into a jar until the jar is about half full. Add a few drops of ink to the water and stir it in.

1

2

Now get another jar and ask the adult who is helping you (your laboratory assistant) to put a few drops (about 6) of household bleach into it. Pour water onto the bleach until that jar, too, is about half full.

3

4

Now pour some of your bleach mixture into the ink mixture. Through the magic of science the ink mixture will become clear!

Bleaches are often used in the house to whiten material and that is what is happening to your ink mixture. Oxygen in the bleach is causing a chemical reaction to the ink to make it clear. Is that clear?

6

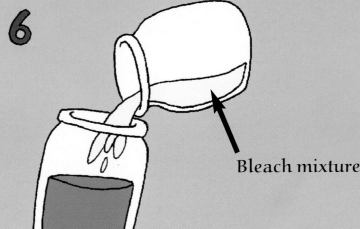

5

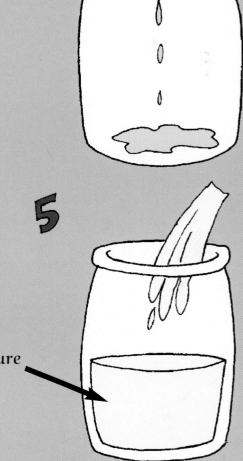

Bleach mixture

61

Magic Circles

Pour some water into a shallow bowl or
saucer. Now very carefully pour a very thin
layer of cooking oil onto the surface of the
water. (You'd better ask the cook's
permission first, as that oil may be needed
for cooking and you could lose out on some
delicious chips or something.)

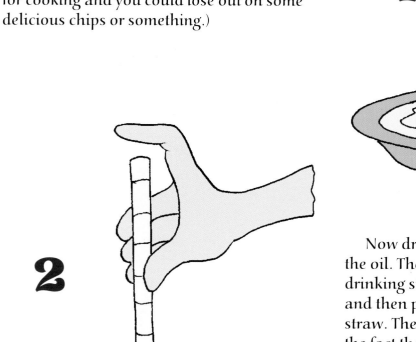

Now drop some small drops of ink onto
the oil. The easiest way to do this is to use a
drinking straw. Dip the straw into the ink
and then put your finger over the top of the
straw. The ink stays in the straw owing to
the fact that your finger stops air pressure
pushing it out (more science). When you
release your finger the air pressure returns
to normal and the ink drops out.

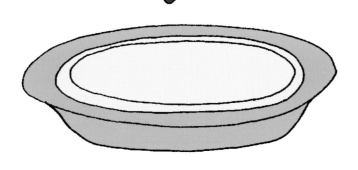

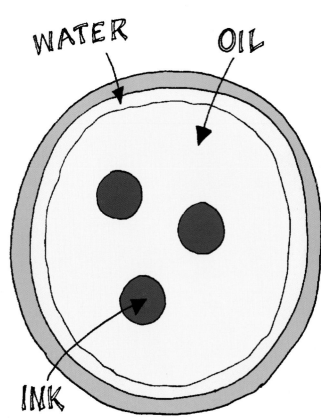

Only allow small drops to come out of
the straw onto the oil. The ink does not
spread out across the surface of the oil as
you might expect but forms into perfect
circles. This is because the molecules in the
oil force the ink to get itself into the smallest
possible area – and that happens to be a
perfect circle. It may look like magic but it is
really science that does the trick.

A Merry Dance

Why did the butterfly get dressed up?
She was going to the moth ball.

In this case it's not the butterflies or moths doing the dancing but the mothballs. In addition to a few mothballs (or even pieces of mothball if they are large), you will need a jar of water, two tablespoonfuls of vinegar, and two tablespoonfuls of bicarbonate of soda.

Get Mixing

Stir the ingredients into the water slowly until all the soda has dissolved. If you do it too fast you will get a very fizzy mixture (there, you always wanted to be a fizzicist).

Drop the mothballs into the liquid and what do they do? They sink! Of course they do! What else did you expect? But have a little patience and the mothballs will start to rise to the top of the water. Having reached the top they will then sink down again. And then they will begin going up and down continuously dancing in a mothball's ball!

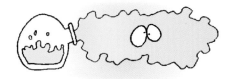

Why?

But what about the serious scientific side of all this? Take a close look at the mothballs and you will be able to see tiny bubbles on them. These bubbles are of carbon dioxide. The bubbles carry the mothballs up to the surface but then they burst into the air so the mothballs lose their buoyancy and sink down again. When they reach the bottom new bubbles form and up they go!

The mothballs will continue to rise and fall until all the chemicals have been used up, which could take several days.

Just one small point to remember – this will not work too well if the mothballs are smooth because the bubbles have nothing to grip onto. If you think they may be too smooth just roughen them a little.

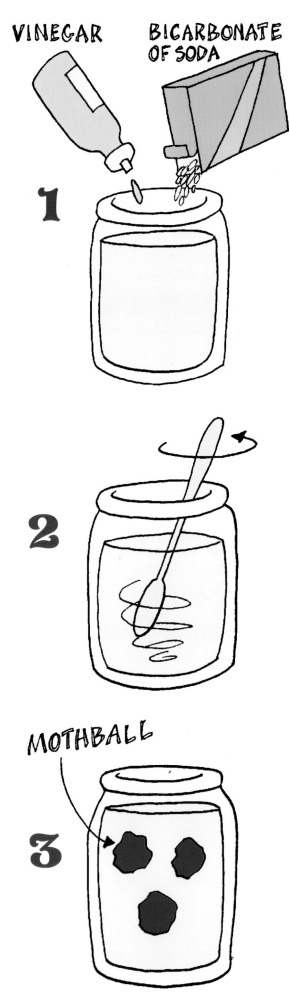

VINEGAR BICARBONATE OF SODA

1

2

MOTHBALL

3

Put Some Sparkle In Your Life

Here's a simple scientific experiment that is guaranteed to put some sparkle in your life.

You will need some powdered rosin. If you play the violin you will probably have some. If you do not play the violin perhaps you can find a friend who does and who is prepared to let you have a piece of rosin. (It's a hard resin used to rub on violin bows.)

Put the rosin in a cloth and hit it with a hammer to pulverize it into a fine powder. Now put the powder into an old pepperpot. Shake the powder over a candle flame and you will produce a lovely shower of brilliant sparks.

1

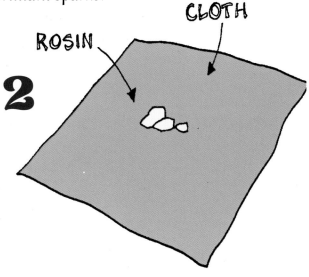

2 CLOTH ROSIN

POUR IN HERE

4

THUMB (DO NOT HIT!)

5

This experiment is quite safe but candles should be handled with care, so get an adult to be your laboratory assistant and help you. What happens is that each grain of rosin burns intensely for just a moment but together they make up thousands of tiny sparks.

64

Not So Rosy Rose

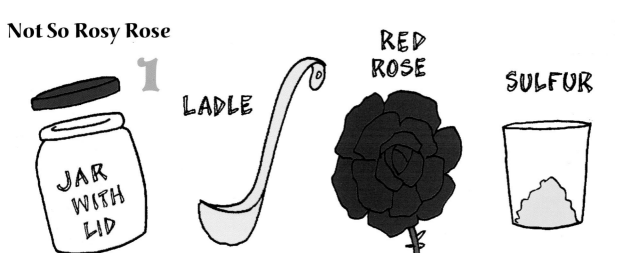

1

JAR WITH LID

LADLE

RED ROSE

SULFUR

Ask your mother if you can borrow a long-handled ladle for this experiment. Put about a teaspoonful of sulfur in the ladle and then hang it in a jar. Put a red rose into the jar and then set light to the sulfur. (If you cannot get any sulfur ask your school science teacher to try this experiment – you may even get extra marks for enthusiasm!)

As soon as the sulfur starts to burn, put the top on the jar. You may be more popular with other people in the house if you do this outside, as burning sulfur pongs like mad.

Within a very short time the red will have turned white. This has happened because in burning the sulfur releases sulfur dioxide and this bleaches the rose.

Make certain that the ladle is washed thoroughly after this experiment, or use something like a large bottle cap to hold the sulfur and then you can throw it away afterwards.

2

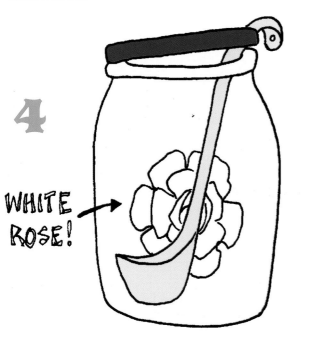

4

WHITE ROSE!

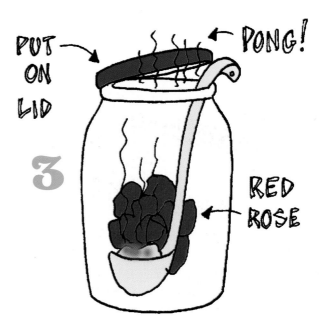

PUT ON LID

PONG!

3

RED ROSE

Bubbles Of Strength

It is very frustrating when you blow a bubble mixture only to have the bubbles burst immediately. But you can use science to make a stronger than normal bubble liquid that will make stronger bubbles that do not burst so easily.

Strong Stuff

Get some washing-up liquid and mix it with a small amount of water. You want it thick and soapy for best results. Now add some glycerine and stir it in quickly until the mixture becomes almost glue-like.

1 WATER

2 GLOOPY!

WASHING UP LIQUID

GLYCERINE

3

Bend the end of a piece of thin wire into a circle and dip this into the bubble mixture. Blow on the mixture until you get a bubble, a bubble that will be stronger than usual thanks to the glycerine.

66

Seven-Fold

How many times can you fold a piece of paper? You would probably say "lots of times" or "as many times as I wanted!" but in fact almost nobody can fold a piece of paper more than seven times!

Try it for yourself. Take any piece of paper and fold it in half. Now fold it in half again, and then again. See how many times you can keep folding it in half. Once you get up to five or six folds it starts to become very difficult. You're probably thinking by now that you could do it much more easily with a bigger piece of paper.

Try it, the result will be exactly the same. What is important is the thickness of the paper. Every time you fold it you are doubling the thickness of paper to be folded the next time. After seven folds it is 128 sheets thick!

SMALL PAPER

1

BIG PAPER

2 FOLD HERE

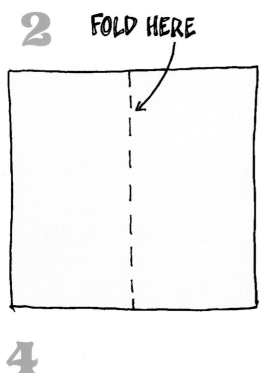

3 FOLD HERE

4

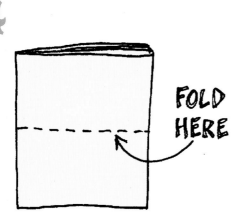

FOLD HERE

Balloon Rocketry

If you take a quick look ahead to page 147 you will see how Newton's laws of motion explain why a balloon rushes around the room when you let the air out. That piece of information can be put to great use in building a homemade rocket. Take an ordinary balloon and tape a drinking straw to it. Thread a 2 meter (about 7 feet) length of thread through the drinking straw. Now set two chairs about 2 meters apart and tie one end of the thread to each one. Make sure the thread is taut. If you now blow up the balloon, pull it to one end of the thread, and let it go, it will rocket between the two chairs. Just make sure nobody has sat down on a chair to read the paper!

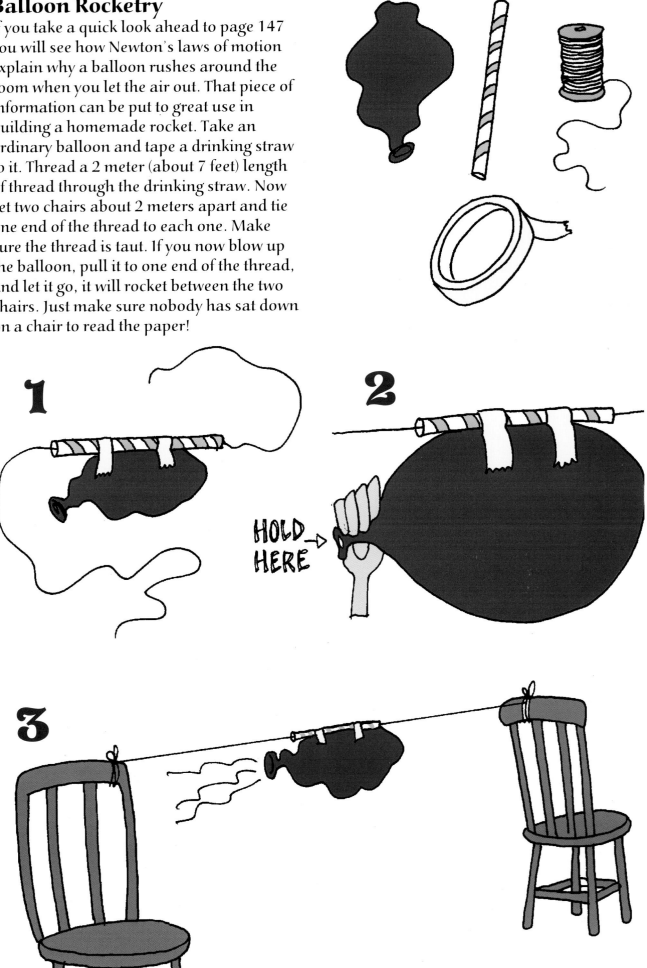

RESULTS

These pages are for you to use!! Every good scientist knows that the result of an experiment – what happens – is the most important part. Very often an experiment goes wrong, or you get a different result from the one you expected. Many great scientists have discovered new and amazing things by doing experiments that went wrong. Perhaps you could be the next one! Write down everything that happened in your experiment – draw pictures too. You never know what you might think of.

Magic Circles

Did the ink drops form perfect circles?

....

How many circles did you make?

....

What happens when two ink circles touch?

....

How can you make the ink circles move?

....

What happens if you stick your finger in one of the ink circles?!

....

A Merry Dance

Did your mothballs rise and sink in the jar?

....

How long did they rise and sink for?

....

When they stopped moving were they at bottom of the jar or floating near the top?

....

What happens if you put more vinegar and bicarbonate of soda in the water?

....

Not So Rosy Rose

Did the rose turn completely white?

....

How long did it take for the rose to turn white?

....

Where did you get sulfur from?

....

What did the horrible pong remind you of?

....

Bubbles of Strength

Were you able to make super strong bubbles?

....

What is the best mixture for super strong bubbles?

....

How big was the largest bubble you made?

....

How long did the most long-lasting bubble last?

....

What does glycerine and washing-up liquid taste like?!

....

Seven-Fold

How many times were you able to fold a piece of paper in half?

....

What is the difference if you use a much bigger piece of paper?

....

What is the difference if you use a much thinner piece of paper, such as a sheet of newspaper?

....

WHOOOSH!

Balloon Rocketry

Did the balloon travel the whole distance between the chairs?

....

How far can you make a balloon rocket travel?

....

What happens if you use a bigger balloon?

....

What happens if you make the thread slippery with oil?

....

Why wouldn't this rocket take you to the Moon?!

....

MISTAKES DEPARTMENT

Professor Knowall, what's the easiest way to become a scientist?

The easiest way to become famous as a scientist or inventor is to make a mistake. It is quite amazing how many things came about by accident. Next time you have an accident don't worry about it. You may be a great scientist one day.

Dyeing To Succeed

William Perkins, an 18-year-old student chemist, was carrying out experiments to make quinine from aniline. The experiment was a complete failure. He ended up with a black, treacle-like gunge. He tried boiling it and, much to his surprise, it turned bright purple!

Purple had always been an expensive dye to obtain, which is why it was used for many royal robes, but Perkins realized that his substance could be used as a dye without too great expense.

Following Perkins' success many more synthetic dyes were developed. Previously dyes had to be made from natural products, but synthetic dyes were easier and cheaper to produce. They also offered a wider range of colors than had been available before.

77

Match Of The Day

John Walker had the surprise of his life in 1827 when he hit a stick against the floor. It burst into flames! It was an illuminating experience and, being a chemist, he realized the fact that the stick had been coated with potassium chloride and antimony sulfide had something to do with it. From this accident, being a bright spark, he developed the first safety match. At first he made cardboard matches but later decided to use pieces of wood.

According to Walker's own records the first person to buy a box of matches was a local solicitor, a Mr Hixon, on 7 April 1827. Walker continued to sell matches as part of his chemist's business at Stockton-on-Tees, County Durham, for the rest of his life.

A Matter Of Safety

The idea of safety glass, of the type used for car windscreens, came about by accident in 1904. A scientist named Bendictus was working in a laboratory in Paris, France, when he accidentally knocked a bottle off a shelf. The bottle smashed but, much to the scientist's surprise, it retained its shape. Bendictus examined the bottle and discovered that it had contained a solution of *collodion* (a syrupy solution of cellulose nitrate in a mixture of alcohol and ether), and this had formed a thin skin of cellulose inside the bottle which held all the fragments together.

Good Year For Goodyear

Although Charles Goodyear experimented for seven years to make rubber usable for making things his actual breakthrough came by accident. Up until 1839 all rubber was useless when cold because it was stiff and brittle. It was not any good when it was warm either because it got sticky and really yucky.

One day Goodyear was rubbing a mixture of rubber and sulfur from his hands when some of it fell on a hot stove. He gathered it up and nailed it up outside the kitchen door in the cold. When he looked at it the following day the rubber was still pliable. By accident he had made what he was seeking. 1853 was certainly a good year for Goodyear.

WHAT COULD I POSSIBLY USE THIS FOR ?

A Mistake Created The Telephone

People who are annoyed by others answering their mobile phones probably think that the invention of the telephone was a great mistake. In fact it was invented as the result of a mistake.

In 1866 Alexander Graham Bell read about some experiments carried out by the German physicist, Hermann von Helmholz. Helmholz had made tuning forks vibrate by passing electricity through them. Bell made the mistake of thinking that the German had transmitted the sound of the human voice and he decided to try to do the same. On 10 March 1876 he achieved success.

AHHA!

Professor Knowall, what were the first words said over the telephone?

Well, you might think that the first word ever said would be "Hello" but it wasn't. The first words said over the telephone were, "Mr. Watson, come here, I want you." They were spoken by Alexander Graham Bell to his assistant on 10 March 1876 – and were the result of an accident: Bell had spilled acid on himself and was calling for help.

Mop It Up

Another thing invented by accident was blotting paper. Sand was used to dry out wet ink until early in the nineteenth century. The change came when a worker in a paper mill in Berkshire, England, forgot to treat some paper with *size* (a gelatin-type solution that puts a glaze on the paper). It was thought that the paper would be no good so the mill owner decided to use it as scrap paper. When he tried writing on a sheet the ink was quickly absorbed into the paper and the idea for blotting paper was born.

New Developments

He tried each of the chemicals in turn on other plates but not one of them developed a picture. Eventually there was just one chemical left. "This has to be the one," he thought as he tested it on a plate. But the plate retained its blank expression – there was nothing in the cupboard that did the trick.

Suddenly he spotted some drops of mercury that he had spilled in the cupboard. It was the vapour from this mercury that had caused his picture to develop. It is said that when he made this discovery Daguerre shouted: "I have seized the light! I have arrested his flight! The sun himself in future shall draw my pictures!" Exactly what you or I would say in the same circumstances!

Some Day My Prints Will Come

In 1837 Louis Daguerre, a French painter, was using a *camera obscura* (a device used to project the image of an object outside onto a screen inside) to help him with his landscape painting. He experimented with plates coated with silver in the hope that he could create an image, but without success. One day he put a used plate in a chemical cupboard, planning to clean it the next day. The following day he was amazed to find that a perfect picture had developed on the plate. The cupboard was full of different chemicals so he reasoned that one or more of them had brought about this magical transformation. His only problem was to find out which.

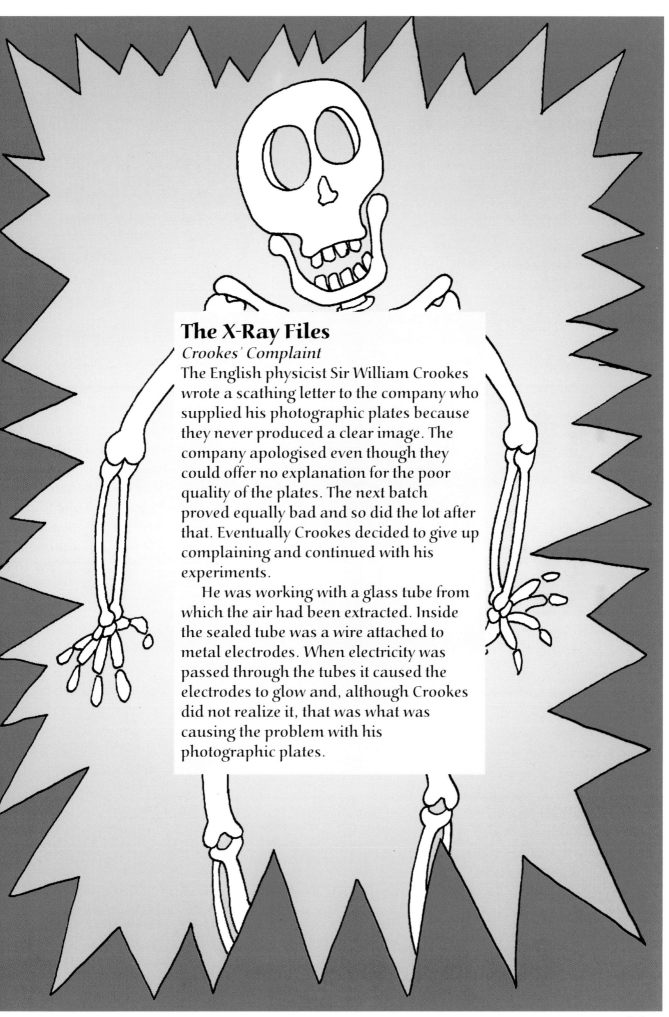

The X-Ray Files

Crookes' Complaint

The English physicist Sir William Crookes wrote a scathing letter to the company who supplied his photographic plates because they never produced a clear image. The company apologised even though they could offer no explanation for the poor quality of the plates. The next batch proved equally bad and so did the lot after that. Eventually Crookes decided to give up complaining and continued with his experiments.

He was working with a glass tube from which the air had been extracted. Inside the sealed tube was a wire attached to metal electrodes. When electricity was passed through the tubes it caused the electrodes to glow and, although Crookes did not realize it, that was what was causing the problem with his photographic plates.

The Bone Show

In 1895 Wilhelm Roentgen, a German physicist, was carrying out some experiment with a Crookes vacuum tube. He discovered that if someone's hand was placed between the tube and a photographic plate the developed plates showed the bones of the hand! He had discovered the reason why Crookes' plates were getting fogged – they were being affected by rays from the tube. Roentgen had discovered X-rays and it was not very long before they were being used in hospitals.

X, The Unknown

The X in X-rays stands for unknown because when the German physicist Wilhelm Roentgen discovered them in 1895 he did not know what they were.

More X-Rays

A year after Roentgen's discovery of X-rays a French scientist, Antoine Becquerel, noticed that some of his photographic paper had gone dark. This had happened in spite of the fact that the paper was stored in a light-proof wrapping.

Then Becquerel saw that the paper had been stored near a chemical containing uranium. When tested it was found that the uranium was giving off mysterious rays, which Becquerel assumed to be the same as those discovered by Roentgen. But Becquerel was wrong – he had discovered radioactivity.

Don't Stick To It

A complete failure by research chemist Spencer Silver resulted in a handy item now found in most offices. Silver worked for the 3M Corporation in America trying to produce the world's strongest glue. In 1970 he tried what he thought was the answer, only to discover that the glue he had produced would not stick permanently to anything. It was a complete and utter failure!

Eleven years later a colleague, Arthur Fry, was getting annoyed because every Sunday he used slips of paper to mark the hymns in his book – and invariably the papers fell out. The ideal solution would be to have a slip of paper that would stick in the book but which could be easily removed. Then he remembered Silver's non-stick glue. The idea was tried, and worked, and as a result what is now known as the Post-It® Note came into being – little slips of gummed paper that can be used to mark papers or books and which can be easily removed.

A LITTLE NOTE TO REMIND YOU

Don't Be Stuck Up

In 1938 Roy Plunkett, an industrial chemist in Wilmington, Delaware, USA accidentally discovered polytetrafluoroethylene. What's so special about polytetrafluoroethylene, apart from its funny name, you may ask. Well, the special thing about polytetrafluoroethylene is that it is not affected by chemicals, sunlight, moisture or heat between -260 degrees Centigrade and 330 degrees Centigrade (-420 degrees Fahrenheit and 626 degrees Fahrenheit). That's equal to the difference between the unbelievable coldness of outer space and the intense heat at whch lead melts. Another special property of (here's that word again) polytetrafluoroethylene is that things don't stick to it. Not even chewing gum will stick to it. Wow, bet you're amazed by that outstanding and very useless bit of information.

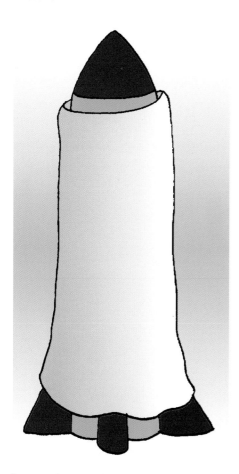

In fact polytetrafluoroethylene (here we go again) has proved very useful stuff. It was first used to handle corrosive chemicals and it later helped protect rockets and missiles from extreme heat.

Keep It Short

Saying "polytetrafluoroethylene" (if you can!) can be quite a time-consuming business so scientists refer to it as PTFE. That makes life as a scientist so much easier.

Plunkett's Pans

In 1958 polytetrafluoroethylene took on a new role, one with which you may be more familiar. The French engineer, Mark Grégoire, used its non-stick properties to coat frying pans. But Grégoire did not use the word "polytetrafluoroethylene" as his brand name (no wonder). He chose "Tefal" which is now famous all over the world. Next time you see someone using a non-stick pan tell them that they owe it all to Roy Plunkett and his accidental discovery of polytetrafluoroethylene!

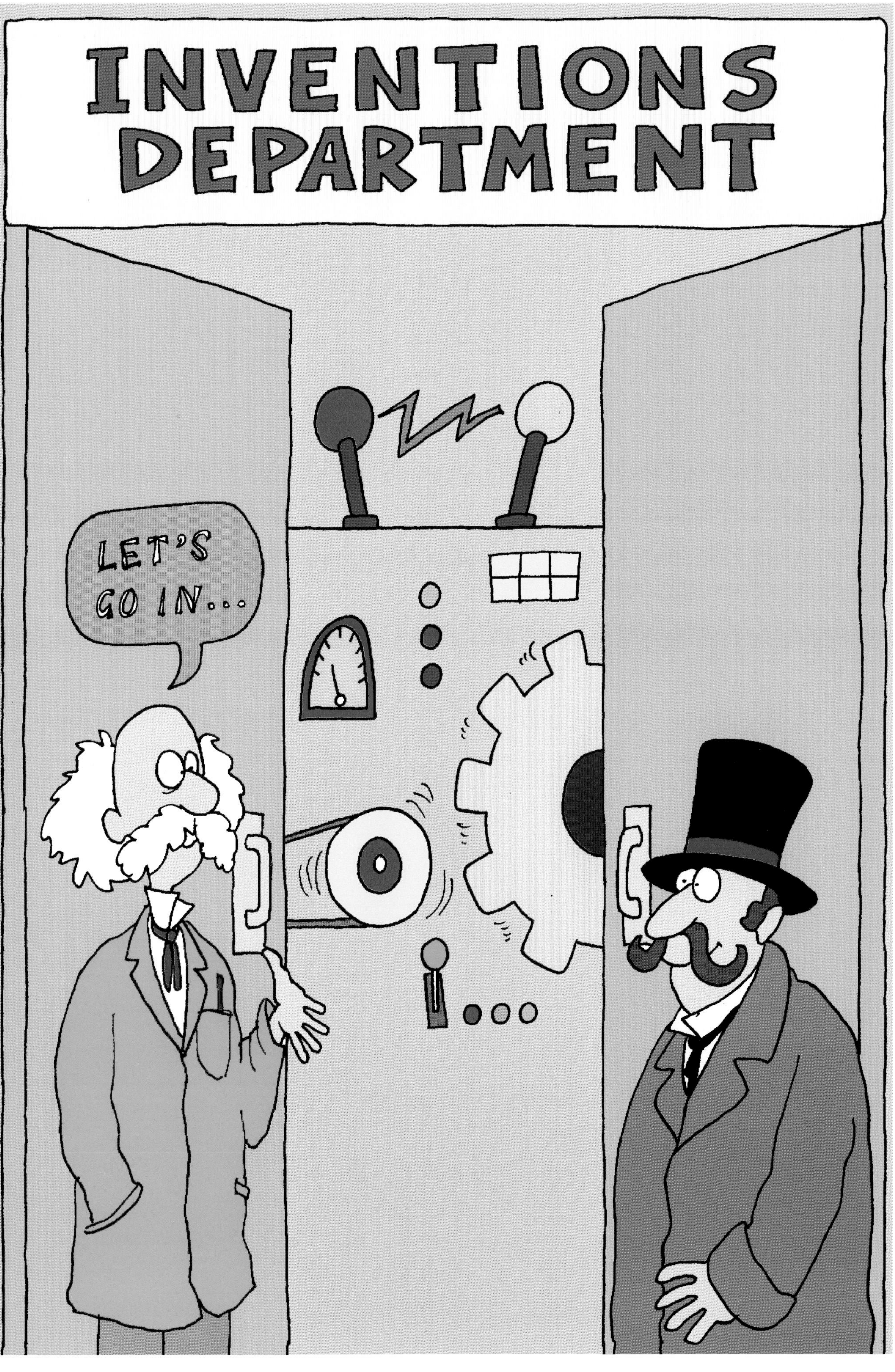
INVENTIONS DEPARTMENT
LET'S GO IN...

INVENTIONS THAT MADE IT

Archimedes' Screw

This remarkable device was devised by the Greek mathematician and inventor Archimedes over two thousand years ago. It consists of a large spiral fitted inside a cylinder. One end of the cylinder is placed in water and then it is turned (the cylinder, not the water). This action pulls the water up the cylinder and out at the top. Archimedes devised it to pump water from the hold of a large ship and later it was used in land irrigation. It is still used in many parts of the world.

Saving Syracuse

In 215 BC the Romans attacked Syracuse on the island of Sicily, which was then settled by the Greeks, but they had not reckoned on the genius of Archimedes, who lived there. He devised a number of ingenious defences that saved the town. These included enormous stone-throwing catapults and a large "claw" which grabbed ships from the water to sink them.

The Sicilian soldiers must, however, have been very surprised when Archimedes asked them to polish their shields as bright as they could. He then used the shields to reflect the sun's rays and set fire to the Roman ships.

The Romans realized they were beaten and laid siege to the city. When they did eventually seize the city in 212 BC Archimedes was, alas, killed by a Roman soldier.

But Does It Work?

Scientists were rather sceptical about stories of how Archimedes burned the Roman ships, so in 1973 they decided to try an experiment. At the Scaramanga naval base near Athens seventy Greek sailors were given oblong mirrors covered with thin sheets of copper. On the word of command each man aimed his "shield" so that it reflected the sun onto a rowing boat near to the shore. Within a few seconds the boat began to smolder and less than two minutes later it burst into flames!

Lever Everything To Me

Archimedes (what, him again!) also developed the theory of levers. The idea had been used before but it was Archimedes who set out the principles of leverage in a scientific fashion.

He is reputed to have said, "Give me a place to stand and I will move the earth." Just as well he never got the chance to try it – if he had we would now be somewhere else in the universe.

Clever Clogs

Leonardo da Vinci, who lived from 1452 to 1519, was the cleverest inventor of his day. He came up with designs for a bicycle, a flying machine, a tank, and a parachute but unfortunately none of them was ever made. Many of his ideas for machines were so far ahead of his time that there was no available means of powering them.

He was a brilliant artist, musician, scientist, mathematician, and biologist. Quite a clever clogs, in fact.

MONA LISA

Genius

"Genius is one per cent inspiration and ninety-nine per cent perspiration."
Thomas Alva Edison (1847–1931)

I'M 99% OF THE WAY TO BEING A GENIUS!

Professor Knowall, have you ever invented anything?

Oh, yes, lots of things. One of my greatest inventions was a pill that was half aspirin and half glue. It was especially for people with splitting headaches.

Getting Up Steam

Around 150 BC the Greek mathematician Hero of Alexandria devised a machine called an *aeolipile* that was powered by steam. It consisted of a ball on an axle above a large water container. When water in the container was heated, steam was forced up two pipes and then into the ball via the axle. The steam came out from two jets on the ball and this made the ball spin.

Hero's device was a form of jet propulsion – now if only he had fixed it to the wing of an airplane the whole history of aviation would have been different.

Pump Action

The first real engine to make use of steam power was invented in 1712 by Thomas Newcomen. It was designed to pump water from a coal mine.

Getting On The Rails

The first steam locomotive to run on rails was the *New Castle*, designed and built in 1804 by the Cornish engineer Richard Trevithick for an ironworks in South Wales. It pulled loads along 16 kilometers (10 miles) of tramway.

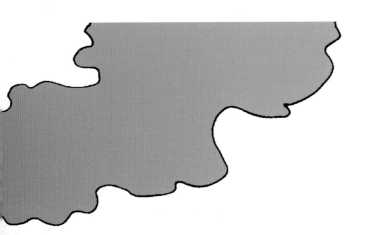

No one Can Live At That Speed!

The British government were very worried about travelling by train. It was believed that anyone travelling at speeds of over 12 miles (19 kilometers) per hour would suffocate.

The First Passengers

On 27 September 1825 the Stockton to Darlington Railway was opened for the transportation of freight. The first train, pulled by George Stephenson's Locomotive No.1, also carried the directors and principal shareholders of the company. The report of this event proved so popular that the directors were forced to make this new form of transport available to the general public, even though that was not the original idea.

Famous Last Words

The German physicist Heinrich Hertz, who discovered radio waves, told Guglielmo Marconi that he was wasting his time trying to send signals by wireless. But in 1895 Marconi succeeded.

TYPICAL!

COMMERCIAL BREAK

Junk Box In The Corner

Just imagine having a pile of junk in the corner of your living room tied together with glue and string, with a hatbox and some knitting needles thrown in for good measure. Doesn't sound much like a television set does it? But that is what John Logie Baird made the first television from in 1924.

YOU'LL HAVE TO TALK TO MY AGENT

Professor Knowall, who was the first person to appear on television?

Well, it wasn't me! It was actually William Taynton, a 15-year-old London office boy, on 2 October 1925. He happened to be passing when John Logie Baird rushed out into the street to find someone. He had just managed to transmit the image of a ventriloquist's dummy and he wanted to try out his equipment with a real person. William was frightened by the bright lights and weird-looking equipment so Baird had to give him some money before he would agree to the experiment.

Rocket Flight

In 1936 the British scientific magazine *Nature* published a review of a book about rocket flight. The reviewer said, "The whole procedure sketched in the present volume presents difficulties of so fundamental a nature that we are forced to dismiss the notion as essentially impractical." But experimental rockets were already being tried out at that time and it was only a few years later that they proved successful.

Powered Flight

In October 1903 the Canadian-born astronomer Professor Simon Newcombe announced that powered flight was "utterly impossible."

Less than two months later Orville Wright became the first man to fly in a power-driven airplane.

The Atomic Bomb
In 1945 Admiral William D. Leahy told the American President Harry S. Truman that the atomic bomb "will never go off, and I speak as an expert on explosives." A few months later the first successful tests of the bomb took place at Alamogordo, New Mexico, USA.

GET IN THE BUNKER!

STUFF AND NONSENSE

Professor Knowall, what was the first liquid fuel rocket?

The first liquid fuel rocket was fired by Robert Goddard near Auburn, New York, on 16 March 1926. For twenty-seven years Goddard had worked towards that moment in the belief that space travel was possible. He did not get very far on that first flight, just 12 meters (39 feet) off the ground, but it was a start.

Who Needs School?

Thomas Alva Edison was America's most prolific inventor but he only had three months' official schooling.

Margarine Wins The Prize

When the French ruler Napoleon III organized a competition to find "a suitable substance to replace butter for the navy and the less prosperous classes" there was just one entry. It came from the chemist Hippolyte Mège-Mouriés, who had been working on just such a project for the previous two years.

Being the only entrant, Mège-Mouriés won the competition! (Spread it around.)

The First Margarine

The first margarine made by Mège-Mouriés was a compound of suet, skim milk, cow's udder, pig's stomach, and bicarbonate of soda. (Sounds revolting.)

Why Margarine?

Hippolyte Mège-Mouriés called his butter substitute margarine because at one point during its production it had the appearance of "a cascade of pearls" and the Greek word for pearl is *margarites*.

Take A Swig Of Brownwigg

The first artificial mineral water was made in 1741 by William Brownwigg, of Whitehaven in north-west England. He found that bubbling carbon dioxide through water gave the water a slightly acid taste, rather like natural mineral waters that were drunk for their health-giving properties.

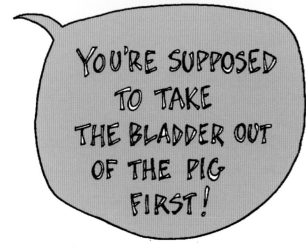

Steam Machine

In 1769 Nicholas-Joseph Cugnot invented the first self-powered vehicle. It was powered by steam and had an enormous kettle-like container on the front. Unfortunately it was very slow and when demonstrated for the French army in 1771 it went out of control and crashed into a wall. Not a very good start for the story of the motor car.

It Tastes Of Pig!

The eighteenth-century English clergyman Joseph Priestley made artificial mineral water as part of his many experiments with gases (he discovered oxygen) but John Nooth, a Scottish physician, said it tasted of pig's bladder. This was because at that time pigs' bladders were used a great deal to hold gases for experiments. Nooth made an improved mineral water by using glass apparatus to make it.

Meat From Car Seats

Robert Boyer, a chemist with the car manufacturer Henry Ford in the 1930s, had the task of searching for a material other than leather to be used for car seat covers. During his research he discovered that the remains of soya beans used in the making of margarine could be spun into strands. As the soya was edible he conceived the idea of using these strands to make a meat substitute.

Boyer's research work meant that he could not proceed with his idea for another twenty years. The spun soya fibres, wound together and then mixed with a flavouring, are now found in many products, especially those prepared for vegetarians. It is called *textured vegetable protein* or, more commonly, *TVP*.

If Boyer had worked for someone called Henry Food rather than Henry Ford no doubt the product would have been made available earlier. On the other hand, if Boyer had made his seat covers from the soya, drivers caught in traffic jams could have eaten the seats if they fancied a quick snack.

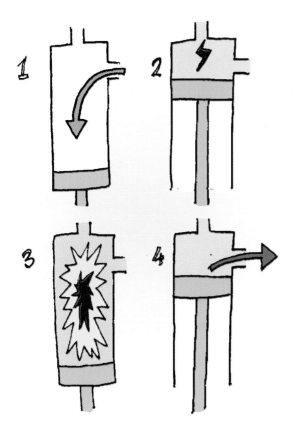

Exploding Cars

Did you know that motor cars are powered by explosions? Perhaps that is why they are often called "bangers"? Gas vapor and air are sucked into a cylinder as the piston moves down inside the cylinder. The inlet valve then shuts and the piston moves back up the cylinder, squashing the gas/air mixture. As the piston reaches the top of the cylinder a spark plug screwed into the top of the cylinder does its job and produces, yes, you've guessed it, a spark. This ignites the gas/air mixture, which burns very rapidly and pushes the piston back down the cylinder.

As the piston moves up the cylinder again an exhaust valve is opened to let the burnt gases out of the cylinder. And then it is back to the start again.

This is known as a *four-stroke cycle*.

Professor Knowall, what does the word "laser" mean?

Well, the word "laser" stands for "Light Amplification by Stimulated Emission of Radiation." Now you know!

Lightning And Electricity

Flash Of Genius

Benjamin Franklin, an American statesman and scientist, proved that lightning was a form of electricity. He did so in a very hazardous way by flying a kite, made from two silk handkerchiefs, in a thunderstorm one day in 1752. He hung a metal key on the kite string and found that, when lightning hit the kite, sparks flew from the key when he held his knuckles near it.

Franklin was lucky he was not killed by the experiment – some others who tried to copy him were.

The experiment did prove useful, however, for it led Franklin to invent the lightning conductor. No that is not a fast ticket collector on a bus, it is a metal rod fixed above the highest point of a building and down into the ground. It is easier for lightning to travel down this rod than through the material of the building so the building remains safe.

A Few Do's And Don'ts When There Is A Storm About

Never ever shelter from a thunderstorm beneath a tree, electricity pylon or similar tall, solitary object. These are probably the most dangerous places you can be.

If you are caught out in the open during a thunderstorm, crouch down as low as you can, preferably not on the highest point of the ground (because if you do, you become the highest point and are liable to be struck by lightning).

If you are swimming when a thunderstorm starts do not stay in the water.

Don't Be A Bright Spark

Electricity plays such an important role in our everyday lives that we often forget that it can be dangerous. Remember these rules:

Never put anything other than a plug in a wall socket.

Never touch any switches or electrical appliances if your hands are wet.

Electrical wires running under carpets are a source of potential danger.

They Must Have Been Quackers

One of the first creatures to take to the air was a duck. It had as its travelling companions a sheep and a rooster. This was in Paris, France, in 1783 and the balloon in which they were carried aloft was one being tested by the Montgolfier brothers, Joseph and Etienne.

Up, Up And Away

The first man to make a balloon ascent was François de Rozier, who went up in a Montgolfier balloon in October 1783. Strange that the Montgolfier brothers did not go up themselves – perhaps they had no confidence in their own invention?

Professor Knowall, who invented radar?

The basic principle of radar, the fact that short radio waves can be reflected, was known in the early 1920s. It was not until the British scientist Robert Watson-Watt begin to investigate it that radar became a practical proposition. The first radar station in the world began operating in Britain in 1935.

Getting A Grip

In 1941 George de Mestral found his trousers covered with burrs that had attached themselves to the material. It is something that has happened to most people who walk in the countryside. But most people just pull them off without thinking any further. De Mestral, on the other hand, did think about the tiny seeds and years later used the idea to produce Velcro – strips of material covered with lots of tiny hooks on one side that cling onto small loops on the other. Velcro is now used as a fastening on a wide range of clothing and other items.

I'M SURE I SHOULD BE USING THIS SOMEWHERE...

Turn Of The Screw

The screw was invented in the sixteenth century. It was a brilliant idea – apart from the fact that the screwdriver was not invented until a hundred years later!

Babbage Was No Cabbage

In the 1830s Charles Babbage designed an "analytical engine" which could carry out complicated mathematical problems. Babbage worked on it for fifty years. It was the world's first computer but unfortunately the technology of that time was not advanced enough to build it.

Semi-Precious Specs

The first glasses were not made of glass. When the explorer Marco Polo visited China in the thirteenth century he found people using lenses ground from semi-precious stones to help them see. So the first glasses were actually made from stone.

> THE GLASSES WORK! I JUST CAN'T GET MY HEAD OFF THE TABLE!

A Real Sucker

We now take the vacuum cleaner for granted but a hundred years ago there was no such thing. A British engineer, Hubert Cecil Booth, changed all that in 1901 when he put a large vacuum pump on a cart and went around houses offering to clean out the dust. A long hose was fed from the cart through the window of the house and the dust sucked out. The machine made so much noise that it frightened passing horses.

The machine really took off later that year when it was discovered that the carpets in Westminster Abbey, where the coronation of King Edward VII was about to take place, were filthy. Booth's machine saved the day and the king was so impressed that he ordered similar machines for Buckingham Palace and Windsor Castle.

> HAS ANYONE SEEN THE CAT?

> YUK!

The Hankie Test

Booth got the idea for his vacuum cleaner in 1901 when he watched a demonstration cleaning of a railway carriage at St Pancras station, London. Dust in the carriage was blown everywhere and Booth decided that suction was the answer. In a restaurant after the demonstration he laid his handkerchief on some upholstery, put his mouth to it, and sucked. The handkerchief turned black with the dirt and no doubt Booth was left with rather a nasty taste in his mouth.

Humpty Dumpty Had A Great Idea

In 1937 Sylvan Goldman, manager of the Humpty Dumpty supermarket in Oklahoma, USA, realized that a shopping basket restricted the amount of goods a person could buy. The basket wasn't big enough and too many goods would be too heavy to carry. He came up with the idea of putting a larger basket on wheels and the cart, or shopping trolley, now seen in every supermarket, was born. What a trolley good idea.

Tin Cans Revolutionized Transport

The first proper model of a hovercraft, made by Christopher Cockerell in 1954, consisted of a cat-food can fixed inside a coffee tin. These were attached to a hairdryer which blew air through the gap between the cans. He found that the air pressure coming from the cans was three time greater than that being put in by the dryer. From this simple experiment the basic principles of the hovercraft were formed.

Going By Tube

The best ideas for inventions are often the simplest. In 1841 an American artist, John Rand, was looking for a better way of storing his paints. After some thought he devised tubes of thin metal that could be rolled up from the bottom. A screw cap on the top prevented the paint from coming out until it was wanted.

This solved Rand's problem and the idea caught on. Today lots of things, such as toothpaste, glue, mustard, and so on, are available in similar squeezy tubes.

Not all inventions are successful. Some just happen to be invented at the wrong time, some prove impractical, and many are just plain crazy.

...AND SOME THAT DIDN'T

Take A Seat

The authorities in Los Angeles, USA were rather concerned about road accidents involving trams. They came up with a device that would actually make it a pleasure to be hit by a tram – a large couch on the front of the vehicle. Instead of being knocked down and killed, they reasoned, the pedestrian who was hit by the tram would simply be scooped up for a comfortable sit-down!

Unfortunately when they tried it out they found that the pedestrian was not scooped up but simply knocked to the ground and injured. When it began to rain and the cushions got rather wet and soggy they decided to abandon the idea.

Tell Me About Your Sewing

Even clever people come up with some crazy ideas. Thomas Edison, who invented the gramophone and electric light bulb, and took out patents on over a thousand other inventions, sometimes got it wrong. One of his failures was a voice-powered sewing machine.

All sewing machines at that time were driven by pedal power. Edison wanted to convert the sound of the voice into energy to power the machine. To give Edison his credit, the device actually worked. The big snag was that the operator had to shout so loud to generate enough power that it proved to be harder work than pedalling – and there was also the risk of ending up hoarse.

Barking Mad?

A Victorian inventor devised a sewing machine that was powered by a dog walking on a treadmill. Now, if the dog had been barking all the time and this invention had been combined with Edison's voice-powered (bark-powered?) machine, it might have proved successful.

Have A Slice Of Pomato

We hear a lot these days about genetically modified foods – where the genes from one plant or creature are added to the genes of something else to make the product last longer or grow better. For many hundreds of years people have been developing new strains of plants by grafting one plant to another or by cross-pollination. In 1977 scientists at a research centre in Holland announced that it had successfully crossed a tomato with a potato. They called it the "pomato" and it was destined to be the latest wonder of the food world. They discovered later that there was just one snag: both the potato and the tomato are related to the deadly nightshade family, and the pomato turned out to be poisonous!

Feathered Flight Of Fancy

In 1865 an unusual design for a flying machine was sent to an American magazine. It consisted of a circular frame around a central latticed tube. The pilot sat in the central tube and ten eagles were attached to the frame. The idea was, of course, that the eagles would take flight and their combined power would be sufficient to lift the man. To control the steering each bird was attached to a set of reins and there were also ten sets of cords to make the birds go up or down. We will never know whether or not it ever got off the ground. It certainly seems very unlikely, but no doubt the inventor thought it was a good idea at the time.

Night-Time Nonsense

Many inventors come up with ideas that they think are OK but which other people think are crazy. In 1980 Fred O'Brien, a graphic designer, came up with a crazy idea which lots of people thought was brilliant. Fed up with all the crazy gadgets appearing in the shops, he decided to invent one that was even crazier – a luminous pocket sundial for use at night.

Fred sent his invention to a magazine specializing in new designs. They thought it was so good that they published a feature on it in their January 1981 issue, despite the fact that the "inventor" had already told them it was just a practical joke.

Much to Fred's surprise, interest in this impossible product increased. The producer of a television programme that featured new inventions asked to know more about it and how it worked. Fred could not resist taking the joke a little further. "It is worked by photosynthetic sound," he said, saying the first crazy thing that came into his head. He added that a Japanese company was thinking of putting it on the market.

Later, when Fred decided that the joke had gone far enough, he telephoned the television producer to say that the Japanese order had fallen through because the angle of the sun in Japan meant that the sundials would tell the wrong time. The television people were very sad that they would not be able to feature this marvellous invention but that was not the end of the story.

Just when he thought the whole charade was over Fred received another telephone call, this time from a businessman who wanted to make the sundials. Fred explained that it was all a joke because, as everyone knows, the sun does not come out at night but the businessman still wanted to go ahead. By March 1981 the luminous pocket sundials were being produced by a factory in Hong Kong.

So, if you decide to invent buttons for a coat of paint or anything else equally silly, let the world know. Someone may decide to make it and you could end up a millionaire!

The Flying Bedstead

In 1953 Rolls-Royce designed a strange-looking contraption to test vertical take-off, the forerunner of today's vertical take-off and landing airplanes. It was nicknamed "The Flying Bedstead" but that name is more deserving of an invention by Dr W.O. Ayers seventy years previously. His flying machine consisted of a metal frame with seven sets of propellers. A tank of compressed air powered four of the propellers and two were powered by foot pedals. These provided the lift and the seventh propeller, powered by a hand crank, provided the propulsion. It is very unlikely that such a contraption would even get off the ground.

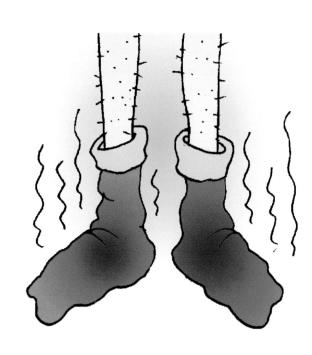

Just In Case

Just in case you were ever shipwrecked a German inventor had the solution in around 1880. It was a suitcase with circular panels in the top and bottom. These could be removed and a large rubber tube with watertight seals fitted through the holes. The owner stepped into this rubber tube and pulled the case up to his waist. The suitcase floated and kept the wearer safe at sea. It might have worked, who knows!

An Explosive Idea

The great inventor Thomas Edison didn't always get it right. He once designed a helicopter powered by gunpowder – and ended up blowing up his laboratory!

Self-Warming Socks

Before turning his attention to the problems of television John Logie Baird tried to make his fortune by making "self-warming" socks.

Street Cred

In 1895 a French inventor came up with a marvellous idea for advertisers. It was a tricycle with messages cut into the tires. On the top of the rear wheel was a tank of printing ink so, as the vehicle moved along, the message on the tyres would be printed on the ground. It seemed a brilliant idea at the time but the street cleaners were not very happy about it, so it was eventually abandoned.

A Head For Pictures

A hat that took photographs was invented by the Luders brothers in 1886. The idea was that it would take candid photographs of people without being noticed. As the photographer had to go under a dark cover to take the pictures it seems very unlikely that such a strange action would go unnoticed!

The Hat In Rhyme

The English magazine *Punch* published a short rhyme about the photographic hat:
*If they knew what I wear when
I walk down the street,
I should be quite a terror to people I meet;
They would fly when they saw me,
and ne'er stop to chat,
For I carry a camera up in my hat.*

Water On The Brain

A German inventor came up with this neat idea. It was a hat designed for tropical wear which had a deep rim to collect rainwater. The water was stored inside the hat and when the wearer wanted a drink all he had to do was take off his hat and turn on a tap inside it.

Some Inventions To Get You Thinking

If you ever fancy the idea of becoming an inventor here are a few possibilities to get you thinking.

Building a snowman takes a long time. How about inventing a large "jelly mould" of a snowman – fill it with snow, turn it upside down and hey presto, you have an instant snowman!

What about a silent alarm clock for people who like to lay in bed late?

Everyone would welcome the invention of a bed that makes itself. Give it some thought next time you are lazing in bed doing nothing.

It might be a good idea to invent a carrier bag that cooks the food on the way home from the supermarket. As soon as you get indoors there is a meal ready and waiting.

We already have chewing gum that is said to clean your teeth. But chocolate that cleans your teeth would be even more popular!

114

Things That Are Unlikely To Be Invented

Inflatable darts board.
Buttons for a coat of paint.
Waterproof tea bags.
Parachutes for submarines.
A solid colander.
Rubber nails.

PSSSST!

BOING!

Death Of The Death Ray

Nikola Tesla, a Croatian inventor born in 1857, once boasted that he could create a death ray. Scientists believed him for, ever since he had set up home in New York, USA, he had been famous for his scientific genius and particularly for the fact that he had been the first to demonstrate how alternating current could be used to provide power over long distances from electricity generating stations.

When buildings began shaking in Manhattan and water mains burst, the police rushed to Tesla's apartment and found him breaking up a strange-looking machine. Tesla told them it was his telegeodynamic oscillator which, although not big, was powerful enough to reduce buildings to rubble.

OOOPS!

When Tesla died in 1942 the American Secret Service searched his belongings for information about this strange device but found nothing. Tesla kept all his scientific findings in his head so to this day no one knows whether or not Tesla did invent some remarkable death ray.

Paper Underpants

Underwear made from paper was hailed as revolutionary in 1965. It was predicted that paper vests and underpants would replace cotton underwear within three years and the idea would be extended to suits and dresses by then. Three years later the idea had died a death.

Look -- No Hands!

It used to be good manners for a gentleman to raise his hat when he saw a lady he knew. But if the man was carrying something this could prove a bit of a problem. James Boyle of Washington, USA, came up with the answer in 1896 – a self-tipping hat. It had a clockwork mechanism inside that the wearer could activate just by nodding his head.

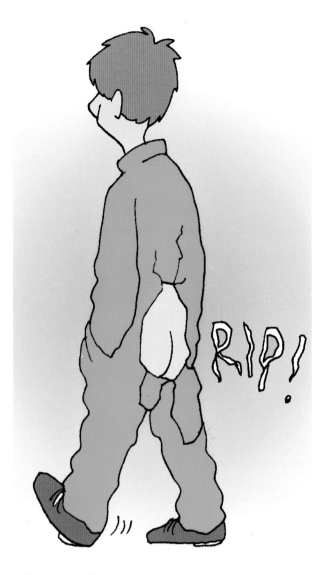

Blotto Idea

Four inventors got together in 1982 and devised an automatic umbrella for rotary clothes lines. When pieces of blotting paper on top of the line got wet the umbrella opened automatically to keep the clothes dry. It seemed like a good idea at the time but no one was interested.

Spiked

It looked just like an ordinary suitcase that any businessman or businesswoman would carry. But if a thief tried to snatch it he would be in for a surprise, for it was a special case dreamed up by English inventor John W. Fisher. At just one press of a button in the handle three 1.8 meter (6 foot) spikes would shoot out of the case, making it very difficult for the would-be thief to get away with it. Apparently it was quite popular during the 1960s but you would be hard pressed to find one now.

GOING ON, AND ON, AND ON, AND ON, AND ......

GRRRRRRR!

I TOLD HIM IT WOULDN'T WORK!

Inventors have always dreamed of achieving perpetual motion: building a machine that, once started, will continue to work forever without using any sort of fuel or external power.

In spite of the fact that scientists say that such a device is impossible, people continue to come up with ideas. So far none of them has worked.

Hammer it out

Around 1245 the French architect Villard de Honnecourt devised a machine that worked with seven large hammers. These were mounted around a large wheel on an axle. According to the Frenchman's beliefs the wheel, once started, would continue to go round and round forever as there would always be more weight on the side of the wheel going down than on the other side.

Magnetic Attraction

Another Frenchman, Pierre de Marcourt, decided that magnets were the answer. He had a hollow wheel with ten *lodestones* (magnets) fixed to the rim. Inside the wheel were a collection of iron nails and another large lodestone. The alternating attraction and repulsion of the magnets would, he thought, keep the wheel moving. But – you've guessed it – it didn't work.

Fludd's Idea Flooded

The seventeenth-century English physician Robert Fludd tried recycling water to power his machine. Water operated a waterwheel which, in addition to providing power for millstones and the like, powered a pump. The pump was used to lift the water back to the top of the *mill race* (the swift current of water powering the pump), and the same water would continue to go round and round.

121

Almost There?

King Charles I of England was very impressed with the machine built by Edward Somerset, Second Marquis of Worcester, in 1638. The machine consisted of an enormous wheel with large balls rolling inside it. It seemed to work – but only actually ran for an hour before coming to a standstill. Somerset continued to work on improvements to his device but never achieved his dream of perpetual motion.

Da Vinci's Disaster

Even the brilliant mind of Leonardo da Vinci was intrigued by the idea of perpetual motion. He also had a wheel but inside his were curved spokes. Between the spokes were heavy balls. As the balls fell to the bottom of the wheel he thought they would provide enough force to lift the following balls up to the top – but they didn't.

An Attractive Idea

In the seventeenth century John Wilkins, Bishop of Chester, devised several perpetual motion machines – or so he thought until he tried them. One ingenious device consisted of a large magnet at the top of a ramp. At the bottom of the ramp was a metal ball, which was attracted up the ramp by the magnet. When the ball reached the top of the ramp it fell through a hole and then rolled back to the beginning – where it was attracted by the magnet again. Sounds like a good idea, but the problem is that any magnet strong enough to pull the ball up the ramp would be too strong to allow it to drop through the hole.

Power From Sponges

A continuous chain of sponges formed the motive power for a perpetual motion machine invented by Sir William Congreve around 1810. Sponges on one side of the chain soaked up water to make them heavy and they were squeezed dry on the other side of the continuous chain. Congreve thought that the difference in weight between the two sides would keep the chain moving. It didn't.

A Load Of Hot Air

In 1856 E.P. Willis of Connecticut, USA, built a superb machine that was on display for several years, and never stopped once. Everyone thought that Willis had cracked it – he had achieved perpetual motion. And then the truth came out. It was all a con: the machine was powered by a hidden system of compressed air.

Water Power

In 1874 people in New York were astounded by the machine built by John Worrell Keely. It was enormously powerful and ran for long periods using just a thimbleful of water as fuel. Keely had great plans for his machine. He was going to build a thirty-car railroad from Philadelphia to New York, which would require just four litres of water as fuel, and a transatlantic liner which would take less than half a litre. His plans never came to anything and when he died in 1898 the reason soon became apparent. The floor of his workshop was torn up and beneath was found the real secret of Keely's amazing machine – an enormous tank of compressed air secretly connected to the machine to make it go.

Alas, it seems that the only perpetual quality of perpetual motion machines is that people perpetually invent them.

Perhaps the nearest man will ever get to perpetual motion is milking a cow, with the milk then flowing to a trough from which the cow is drinking milk!

Scientists say that true perpetual motion is impossible because the laws of thermodynamics state that, whatever form of energy is put into a machine, the amount of energy produced cannot be greater than that put in. In actual fact there will always be a loss of energy. Now you know.

LET'S EXPERIMENT

Let's Experiment
Scientists test their theories by setting up experiments that should be capable of being repeated with similar results. One of the first scientists to adopt this approach was Galileo Galilei in the sixteenth century.

...SOME MORE

Science Can Be Dangerous

In the 1840s the German chemist Robert Bunsen tried making *cacodyls*. These, in case you didn't already know, are bad-smelling poisonous liquids which have the unfortunate habit of bursting into flames without warning. Bunsen learned this the hard way – he lost an eye in an explosion and developed arsenic poisoning which resulted in severe stomach cramps, uncontrollable diarrhea, and partial paralysis.

The experiments given in this section are not so dangerous!

I Know That Name!

If the mention of Robert Bunsen had started bells ringing in your brain because you thought you had heard the name before, you are probably right. The Bunsen burner used in laboratories is named after him. He did not invent it, however. The credit for this goes to Bunsen's assistant Peter Desdega, who actually developed it from an earlier type of burner.

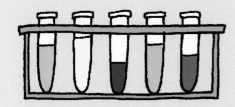

Would You Believe It?

Cold water is heavier than hot water and hot water freezes faster than cold water. Strange, isn't it!

Bouncing Light

You cannot bounce light in the same way that you can bounce a ball, but you can bounce it. All you need is a mirror and a torch.

Put the mirror on the floor in front of you. Now switch on the torch and point it at the mirror. The light will bounce off the surface of the mirror and will be reflected on the wall or the ceiling. Where the light falls depends upon the angle at which you are pointing the torch towards the mirror, for the light will be reflected from the mirror at the same angle that it hits it.

You will get best results if you do this in a slightly darkened room.

You Can Bend Light

You do not have to be a strongman to bend light. It is in fact very easy to do. Fill a glass tumbler with water and put a spoon in it. Look at the spoon from the side and it appears to be broken at the point where it enters the water.

This is due to the fact that light travels faster through the air than it does through denser material such as the water and the glass. This difference in appearance is called *refraction*. It is for the same scientific reason that a river looks shallower than it really is, a thing to bear in mind if you ever go paddling.

Yet another example of the same phenomenon can be seen if you have ever dived to the bottom of a swimming pool to get something. You will know that the object is never in the position you thought it was in when you saw it from above the surface.

Make A Thermometer

A thermometer, in case you do not know it, measures temperature. The name comes from two Greek words, *therme* meaning "heat" and *metra* meaning "measure" (just thought you would like to know that).

You can make a thermometer to measure air temperature from a milk bottle, a drinking straw, and a few other odds and ends.

1

COLORED WATER

First fill the bottle with colored water, having drunk the milk of course. Put the drinking straw into the bottle and use a piece of modelling clay to hold it upright.

Glue a piece of white card to the straw. This will help you see any changes in the water level.

2

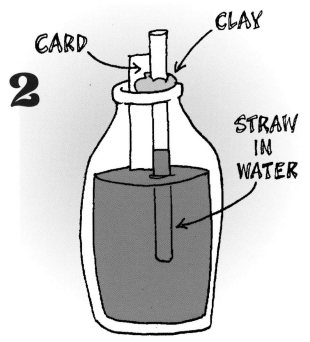

CARD   CLAY

STRAW IN WATER

If the air temperature goes up the water level in the straw will go up. In colder weather it will go down. So you will know whether it is warm or cold – another way of telling is, of course, that if you are sweating it is warm and if you are shivering it is cold.

Under Pressure

Stretch a piece of rubber balloon over the top of a jam jar. Keep it in place with a strong elastic band. Put a dollop of glue in the centre of the balloon and press a drinking straw onto it. (Read the instructions on the glue before you do this as some glues will dissolve the rubber of the balloon.)

Place a piece of card behind the free end of the straw so you can measure its movement.

As air pressure presses on the surface of the balloon the end of the straw will go up – indicating high air pressure. When the pressure decreases the straw will go back down.

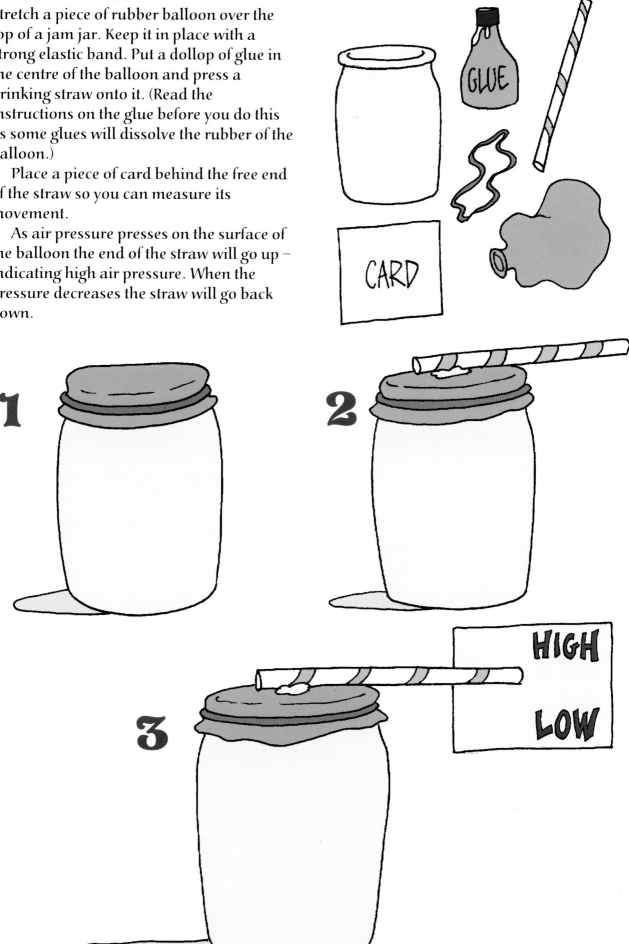

Lever power

With just two broom handles and a long piece of rope you can demonstrate the power of leverage. Tie one end of the rope around one end of one stick and then wind it around the sticks as in the picture.

Get a friend to hold one stick and one more to hold the other. If you haven't got two friends try another experiment, or go home and watch television.

Ask your helpers to hold the brooms apart while you exert your superhuman muscle power to pull them together. Pull on the loose end of the rope and you should find it quite an easy matter to pull the brooms together, no matter how strong your friends are.

Pressure Points

Even water exerts a pressure, and the deeper the water the greater the pressure. You can show this with a simple experiment. All you need is a long container – an old washing-up liquid bottle will do if you cut off the top (make sure it is empty first).

Make three small holes at equal distances from each other down one side.

Now fill the container with water. Naturally the water will come out through the holes you have made, but because the pressure is greater at the bottom, the water from the bottom hole will spray out the furthest and that at the top (where there is less pressure) will not flow out quite so far.

Color Sense

The great British scientist Isaac Newton split colors with a prism. The colors formed a band of strips of different colours, each blending into the next. This is known as a spectrum.

When rainbows form, droplets of water act as prisms to separate the colors. In order to see a rainbow the sun has to be shining, it has to be raining, and you have to be in the middle with the sun behind you.

Try It!
You can do the opposite and make all the colors of the rainbow into white with this experiment.

You will need a disc of card. With a pencil divide the disc into six equal portions and then color each segment one of the colors of the rainbow – red, orange, yellow, green, blue, and violet.

Push a pencil through the center of the disc and then spin it like a top. If it is spun fast enough all the colors will merge into one and the disc will appear to be white. If the disc has a tinge of color when you spin it you probably have too much of that color on the disc. Try making another one with a little less of that color on it.

Naturally – and you might have guessed this – this disc is known as *Newton's Color Disc.*

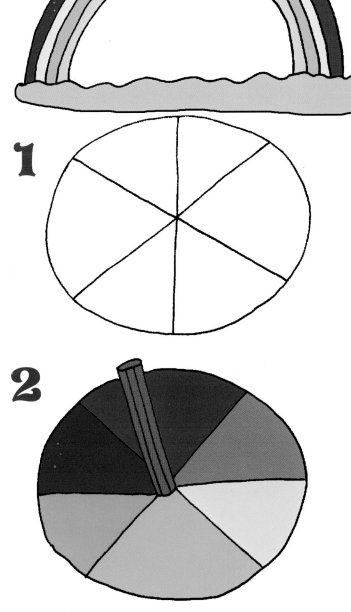

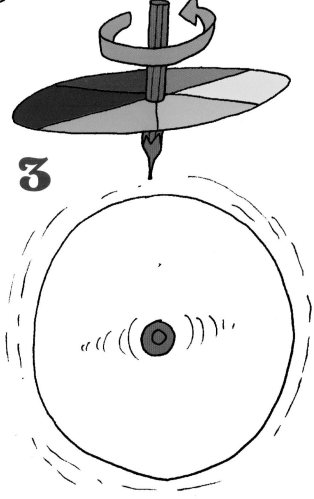

Make A Rainbow

It is not always easy to get hold of a prism but you can still make your own rainbow. All you need is a sunny day, a glass of water, and a sheet of white paper.

Place the glass of water just inside a window through which sunlight is coming. Place the paper on the floor. The water acts just like a prism and will produce a rainbow effect on the paper.

Thales Was No Sap

Our word "electricity" comes from the Greek word "elektron," which means "amber." You may wonder why this is so (all scientists wonder why – that is why they are scientists). The reason is that amber when rubbed will attract small pieces of wool, straw or paper. Rubbing causes the amber to become charged with what is now known as static electricity.

This was first discovered over 2,500 years ago when the Greek philosopher Thales was studying a piece of amber, which is in fact fossilized tree sap. He discovered that when it was rubbed, small bits of feather and cloth stuck to it.

You can do the same thing with glass and certain plastics.

Charge!

To make an electric charge rub a balloon with a piece of wool. This should generate enough electricity to allow you to stick the balloon to a wall.

Water Defies Gravity

Air pressure is actually greater than water pressure and you can prove it with this experiment.

Fill a glass tumbler right to the top with water. Place a piece of card over the glass, being careful not to spill any water as you do so. You must also be careful that there is no air trapped in the glass.

Press the palm of your hand on the card and then turn the glass upside down. Oops! Maybe you should have been warned to do this over a sink as it may take a little practice to get it right.

Remove your hand from the card and the water will defy gravity and remain in place.

Air pressure pushing up on the card keeps it in place. There is pressure from the water pushing down on the card but this is less than the air pressure.

The card must, of course be bigger than the mouth of the glass or air could seep in and press down on the card, and you will end up with wet feet!

Ain't science wonderful!

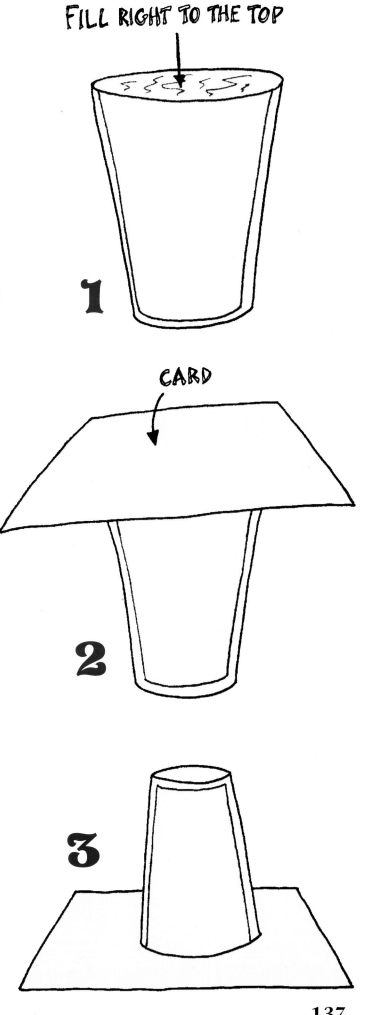

FILL RIGHT TO THE TOP

1

CARD

2

3

Eureka Streaker

The principle named after the Greek mathematician Archimedes states that when something (or someone) is weighed in air, and then in water, there is a difference in weight equal to the amount of water that is displaced.

It is said that Archimedes came to this realization when he stepped into a full bath one day and the water overflowed. He was reputedly so excited by this discovery that he ran naked into the street shouting, "Eureka!" (I have found it!)

No doubt any little old lady would have said, "Yes, dear, you do reek-a, you haven't had your bath yet."

The Dodgy Crown

The Greek king, Hieron II, asked Archimedes if his new crown was made of pure gold. Archimedes used his principle to test this.

First he weighed Hieron's crown, then he measured how much weight it displaced. This was equal to the crown's volume. He then found out the volume of a piece of pure gold of the same weight. It turned out to be less than that of the crown, proving that the king's new headgear was not made of pure gold.

DON'T TRY THIS AT HOME!!

HMMMM...

EUREKA!

Testing Archimedes

You can test Archimedes' principle for yourself and you do not have to run naked through the streets to do it.

All you need are two jugs and an apple. One of the jugs should be filled to the brim with water. The smaller of the jugs, which must have measures on it, is placed beneath the spout of the first.

Now carefully put the apple into the large jug. Water will be displaced and pour into the smaller jug. The weight of the water that has overflowed will be the same as the weight of the apple in the water.

1 FILL TO THE BRIM WITH WATER

2

How Displacement Works

If an object weighs, say, 60 grams (2.1 oz) in the air it may weigh only 40 grams (1.4 oz) in water (because the water supports it), the difference being 20 grams (0.7 oz). In placing the object in the water, 20 grams (0.7 oz) of water will be displaced.

The same is true of something that floats. If a piece of wood that weighs (say) 15 grams (0.5 oz) in the air is floated in water it will weigh nothing. The difference of 15 grams, is the amount of water that it will displace.

Spinning Spiral

You can show that hot air rises with just a piece of card and a length of thread. Draw a spiral on the card and then cut along the lines. Pull the spiral out by pulling the ends apart slightly. Now attach a piece of cotton to one end.

Hold the apparatus above a warm radiator or a table lamp. Try to hold it as still as possible. As the air rises from the heat source it runs around the spiral and will cause it to spin slowly.

1

2

3

4

140

Ice Magic

Using some simple science you can do an amazing magic trick. Show your friends a cup of water in which an ice cube is floating. Also on the table or nearby you should have some table salt and a length of cotton or thin string.

Ask your friends if they can lift the ice cube out of the water with the string. They are not allowed to use their fingers or any cutlery to get the cube out.

When your friends admit defeat you simply rest part of the string on top of the ice cube. Now sprinkle some salt over the cube and wait for a few seconds. Lift the other end of the string and the ice cube will be lifted from the water, sticking to the string.

The reason this works is that the salt lowers the freezing point of the surface of the ice cube so it melts. When the water has dissolved the salt, the ice cube freezes over again, with the string embedded in the ice firmly enough to support the weight of the cube.

The Science In A Snow Ball

Bet you never thought that making a snowball was a scientific experiment, but it is. When you pick up the snow and press it between your hands you are lowering its melting point on the outside of the ball. When you open your hands the melting point of 0 degrees Centigrade (32 degrees Fahrenheit) is reinstated. The surface refreezes and forms a thin skin of ice which holds the snowball together until it hits your enemy.

So the next time your mother tells you off for coming into the house covered in snow tell her that you were carrying out a

PRESSURE MELTS SNOW

MELTED SNOW FREEZES HARD

THESE SCIENCE EXPERIMENTS ARE GETTING OUT OF HAND

142

Snow Business

You need some snow on the ground for this experiment, one of many conducted by the great American statesman Benjamin Franklin. Put two pieces of cloth, one light-colored and one dark, on the surface of the snow. Leave them there for as long as your impatience allows.

When you lift the cloths you should find that the snow beneath the dark cloth has melted more than that under the light cloth.

This happens because light material reflects the sun's heat, so it stays reasonably cool underneath, whereas dark colors absorb the heat and the snow melts.

It is for this simple reason that people in very hot countries usually wear light-colored clothing.

Making Money

Put a coin in the bottom of a cup and then stand so you cannot see the coin.
Now ask someone to pour some water into the cup. The coin appears by magic.
No, it's not magic it's refraction.

Let's Go To The Flicks

You can see how film works by making your own flicker book.

Get an old book or a thick magazine. Starting at the front of the book draw an image – a stick figure is the easiest to start with. On each successive page draw the same image, in the same position on the page, but advance the movement just a little.

Try doing a man running to start with – the first picture shows him still, the next with one leg moved forward slightly, the next with the leg further forward, and keep moving the leg until it is as far forward as it will go without the poor chap screaming out in agony. Continue to draw the pictures, one on each page, with the other leg moving forward. Go all the way through the book like this (or until you get fed up with it, whichever is the sooner).

If you now use your thumb to flick through the pages you will see the man running. A movie works in exactly the same way – it is just a series of still pictures, each moving the action on just a little each time.

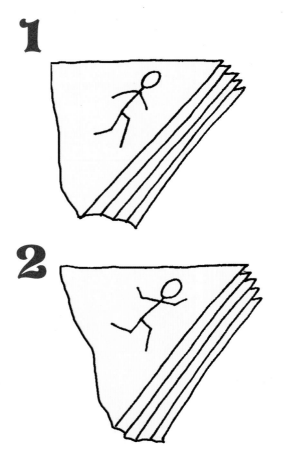

1

2

KEEP GOING!
(IT'S WORTH IT)

USE YOUR THUMB TO FLICK THROUGH

3

ACTION!

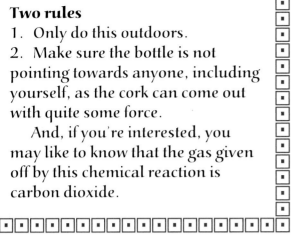

Gas Gun

Put a tablespoon of bicarbonate of soda in a plastic drinks bottle (the bottle must be dry). Now pour in a cup of water. Next put in a tablespoon of vinegar.

 Now you must act quickly, so it is a good idea to have everything ready in advance. As fast as you can, put a cork into the neck of the bottle (this must be a good fit). Lean the bottle up against some stones and all of a sudden the cork will be blown out with a bang.

145

Hard Air

For this experiment you will need a few drinking straws and a potato.

First try to push one of the straws into the spud. The straw is not strong enough for this task and will simply buckle up. You probably expected that to happen anyway.

Now hold the potato firmly in one hand and take a fresh straw in the other. Put your finger over the top of the straw. Stab the straw quickly into the potato. Keep your finger over the top of the straw as you do this.

With practice you will be able to stick the straw right into the potato without bending it (the straw, not the spud). You may even be able to make it go right though the potato.

The reason this happens is that your finger traps a column of air inside the straw and this gives the straw extra strength and rigidity.

What's In A Name?

Many ordinary everyday substances have complicated chemical names. Here are just some of them:

antifreeze	thylene glycol
caustic soda	sodium hydroxide
chalk	sodium carbonate
common salt	sodium chloride
cream of tartar	potassium bitartrate
Epsom salts	magnesium sulphate
plaster of Paris	calcium sulphate
talcum powder	hydrated magnesium silicate
vaseline	petrolatum
vinegar	dilute acetic acid
washing soda	crystalline sodium carbonate

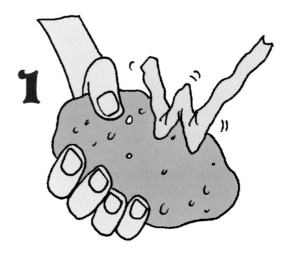

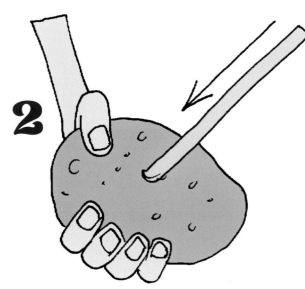

Professor: I'd like a packet of acetysalicylic acid tablets, please.
Pharmacist: You mean aspirin?
Professor: Oh, yes. I can never remember what it's called.

Newton's Laws Of Motion

In 1687 the British scientist Sir Isaac Newton published his three laws of motion, which are:

1. A body resists changes to its state of motion. So something at rest remains at rest until some external force causes it to move. (Like you lazing in bed until someone tips you out of it.)

2. The force to make an object accelerate depends upon the mass of the object and the strength of the force. (It's harder to stop a cannonball than a ping-pong ball going at the same speed because the cannonball is heavier.)

3. For every action there is an equal and opposite reaction. (Blow up a balloon and let it go – the air rushing out of the neck of the balloon makes the balloon go forwards.)

Moving Forward

You can test Isaac Newton's law that states that for every action there is an opposite and equal action quite easily.

From a sheet of card cut out a boat shape. From the stern of your boat cut a channel leading to a small hole.

Lay your boat on some water. Now pour some washing-up liquid or some oil into the hole. Normally the oil would spread out across the surface of the water but the card will stop it from doing this. So the oil is forced out through the channel. The force of the oil pouring out from this channel will cause the opposite reaction of making the boat move forwards.

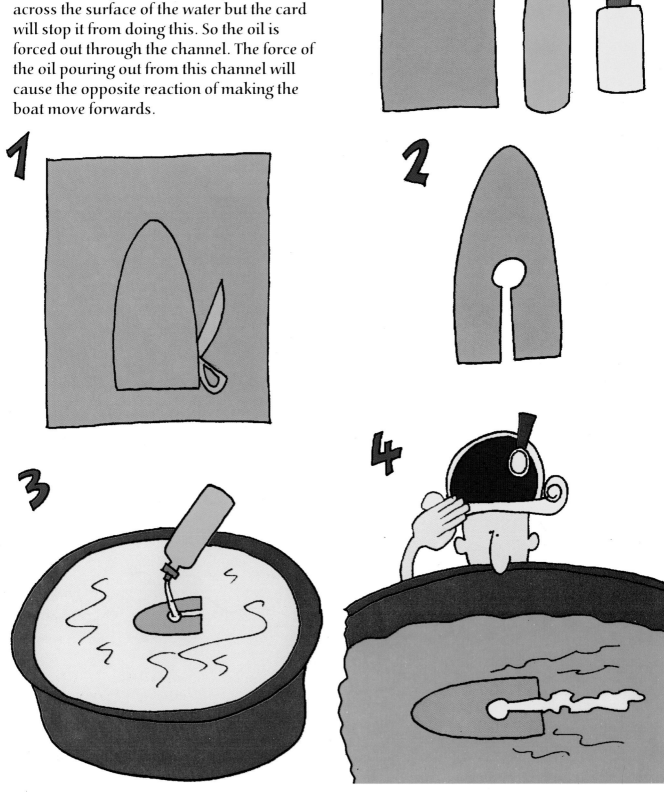

1

2

3

4

148

Taking the plunge

Push two sink plungers together, mouth to mouth (if sink plungers have mouths), as hard as you can. Now try to pull them apart. You will find this extremely difficult, if not impossible, and you may have to get someone else to help you.

The reason the plungers hold together so obstinately is that when you pushed them together you squeezed out most of the air. All the air around us is exerting a pressure (somewhat amazingly this is called *air pressure*). This pressure is being exerted equally in all directions – up, down, sideways, etc.

When you removed the air from the sink plungers you reduced the air pressure inside. So the air pressure on the outside is pushing harder against the plungers than the air pressure on the inside pushing against it.

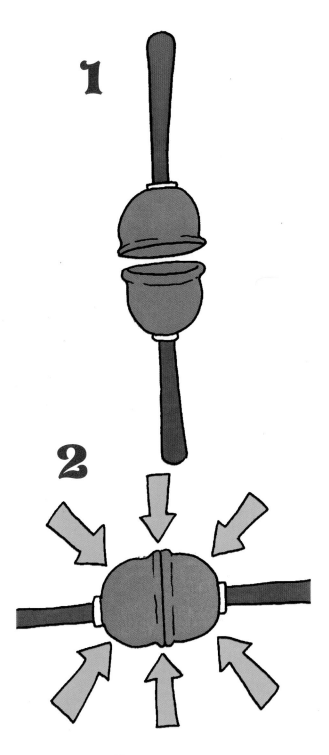

A Hard Pull

In 1654 Otto von Guericke amazed everyone with the sink plunger experiment but he didn't use sink plungers! Instead he had a large hollow copper globe made. This was cut in half and the two halves put back together again. All the air was then pumped out of the globe. The pressure of the air on the outside was so great that it took two teams of horses to pull them apart. This experiment is now known as the Magdeburg experiment after the German town where it took place.

The Descartes Diver

The Descartes diver, named after the seventeenth-century French scientist René Descartes, consists of a hollow figure of a man which is floated in water contained in a long glass jar. A sheet of rubber is stretched over the top of the jar and tied in place. When the rubber is pressed the diver sinks to the bottom of the jar and when it is released he floats to the top again.

The reason this happens is that pressing the rubber increases the water pressure inside the jar. This forces water into the diver through a small hole in his foot. The additional weight causes the diver to sink to the bottom of the jar. When the pressure is released the water flows out from the diver and he goes back up to the top.

Take A Dive

All you need to try out Descartes' diver, more often called a *Cartesian diver* (from Cartesius, the Latin name for Descartes), for yourself is a glass, or jam jar, full of water and a very small bottle. Put the bottle, minus its top, upside down into the water. Allow just enough water to flow into the bottle so that it will still float upside down in the water. If need be, top up the glass with water.

Cover the top of the glass with your hand. Make sure your hand completely covers the whole area. The bottle should now sink down to the bottom.

When you remove your hand the bottle will go back up again.

If you want to you can make a rubber top for the jar and hold it in place with an elastic band. Provided the jar is full and the cover is airtight, this will work extremely well.

150

Core!

One day in 1666 Isaac Newton was sitting in his garden at Woolsthorpe Manor, near Grantham, England, when an apple fell on his head. Newton did not say "Core!" as one would expect because he was sitting beneath an apple tree at the time so there was nothing remarkable about the occurrence – until the great man started thinking about it.

If you drop something it falls to the ground. Why doesn't it fall up, or even sideways? Newton's ponderings eventually led him to the conclusion that all matter attracts other matter and that gravity affects not only things on Earth but even the way that planets orbit the Sun. He had worked out the law of gravity.

Imagine what it would be like without gravity – especially for the people on the other side of the world!

Sign, Please

The great physicist Isaac Newton went to school in Grantham, Lincolnshire, England. One of the many things he did while he was there was to carve his name in a windowsill. It is still there to this day. Not only did Newton discover the law of gravity, he also discovered the law of graffiti!

Seeing Sounds

Seeing sound! That's impossible! Sound is invisible so it stands to reason that you can't see it.

A good scientist, however, never takes anything for granted. Someone says you can't see sound so the scientist asks, "Why not?"

Sound actually consists of waves or vibrations in the air. These waves are just like the ripples made in a pond when a stone is thrown in. Just as you can see the ripples, so with this experiment you will be able to see the sound waves.

First you will need a tube. This could be a tin can with both top and bottom removed (get an adult to do this neatly for you, as you don't want to cut yourself), a plastic cup or cardboard food carton with the bottom cut off, a piece of plastic piping, a megaphone, or anything similar. The main thing is that it must be bigger than your mouth. (Oh gosh, where can you find something that big?)

Over one end of the tube place a sheet of rubber or cling film. Hold it in place with an elastic band.

Use a small piece of sticky tape or glue to stick a piece of aluminium foil (or a very tiny mirror) onto the rubber.

Hold your amazing sound tube with the mirror facing the sun so it is reflected on a wall nearby. If it is not very sunny you could darken the room and get someone to shine a torch onto the foil. Now talk into the open end of the tube and watch the reflection of the mirror. It will be bouncing up and down as you speak.

The reason this happens is that sound waves from your mouth cause the rubber to vibrate and these vibrations move the foil.

LOOK HERE

SHINE LIGHT HERE

SHOUT HERE

Common Sense

T. Huxley once said, "Science is nothing but trained and organized common sense."

String-Along-A-Spoon

Tie a loose knot in the center of a fairly long piece of string (about 1$\frac{1}{2}$ meters, or 5 feet if you prefer). Push the handle of a spoon into the knot and then tighten the knot.

Press the ends of the string to your ears with the spoon dangling down in front of you like some official chain of office (Your Royal Worship the Mayor of Spoon).

Swing the spoon so it hits a table top or the back of a wooden chair. The sound you hear will be pretty impressive – more like a bell than a spoon! That's because solid objects (such as string) are good at carrying sound waves, while air (through which you normally hear sound) is not so good.

When you have got fed up with listening to that spoon, slide it out of the knot and put in a spoon of a different size to see what that sounds like. If you really are a glutton for punishment you could try hanging other things from the string to see what they sound like.

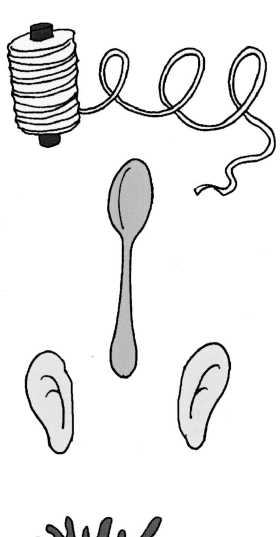

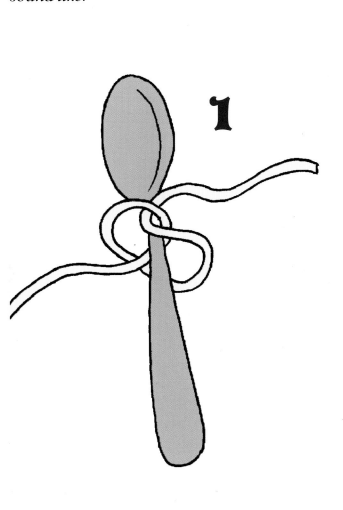

Tower Of Inertia

According to science a thing that is stationary will continue to remain stationary until someone or something moves it. This resistance to movement is called *inertia*. Science also states that something that is moving will continue to move until someone or something stops it. This, too, is inertia.

An example of the first type of inertia is you lazing in bed, too lazy to move, until you are forced to do so. A better example is a motor car. It takes more energy to get it going than to keep it going once it is moving.

Once you are out of bed you can try this experiment. Build a tower of about ten counters from a game of checkers (also called "draughts") and place another, single, counter, about $2\frac{1}{2}$ centimeters (1 inch) from the tower.

Now give a good, hard snap with your middle finger against the single counter (if you have delicate fingernails hit it with a ruler) so it hits the counter at the base of the tower. With a bit of luck the bottom counter will be knocked out from beneath the tower but the rest of the tower will remain standing.

This is simply due to the fact that inertia keeps the rest of the tower from moving.

More Inertia

While you have that tower of counters still standing you can try this. Get a ruler and hit it sharply against one of the middle counters. The middle counter should fly out of position, the rest of the tower remaining intact.

With practice you may be able to knock out any of the counters without disturbing the rest.

INERTIA IN ACTION!

ZZZZZ

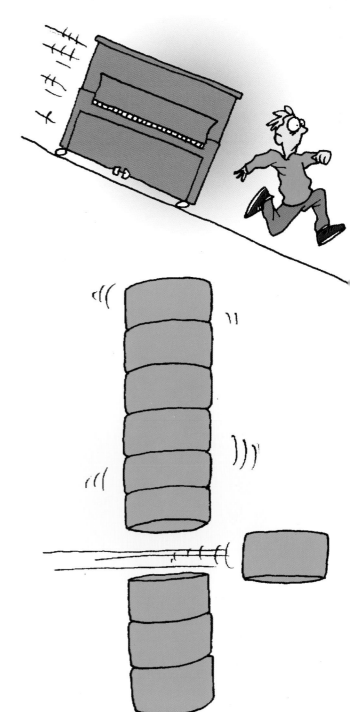

Spud Balance

Here is an amazing looking experiment you can try. Push a long pencil through a potato and then stick two forks into the potato, one on each side. Stand the point of the pencil on the base of a tumbler and you will see it balance at quite an unusual angle. You can change this angle by moving the forks up or down.

What is happening is that you are changing the centre of gravity of the whole lot by moving the forks. You know, of course, that gravity is the attraction between an object and the Earth. The center of gravity is that point in something where there is an equal weight on each side. It's a bit like you putting your arms out to balance yourself when walking along the top of a wall, or a tightrope walker using a long pole to keep an equal weight on both sides of the body.

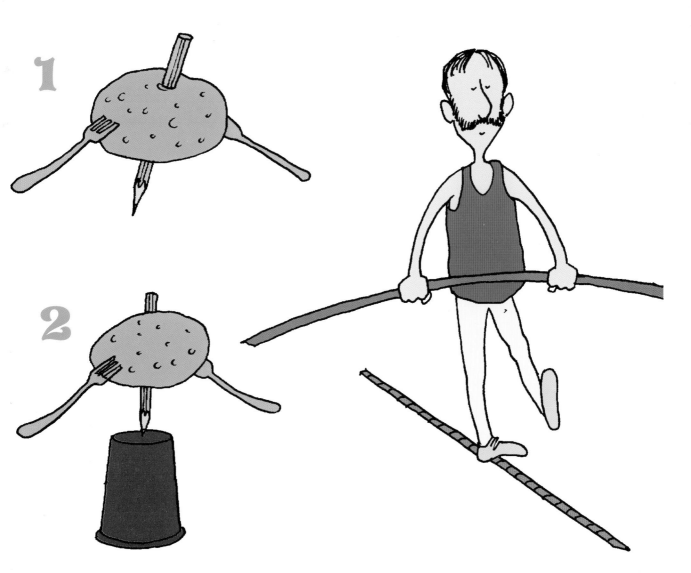

1

2

155

RESULTS

These pages are for you to use!! Every good scientist knows that the result of an experiment – what happens – is the most important part. Very often an experiment goes wrong, or you get a different result from the one you expected. Many great scientists have discovered new and amazing things by doing experiments that went wrong. Perhaps you could be the next one! Write down everything that happened in your experiment – draw pictures too. You never know what you might think of.

Pressure Points

How far did the water go?

....

Which hole squirted water the furthest?

....

What happens if you make the holes bigger or smaller?

....

How many times did you have to change your socks?!

....

Color Sense

Did your colored disc look white when it was spun?

....

If not, why not?

....

Is it better to spin the disc quickly or slowly?

....

Is it better to use dark colors or light colors?

....

Did you get dizzy?

....

Make a Rainbow

Did you see a rainbow?

....

Where is the best place to put the glass of water?

....

Can you produce the same effect using a torch instead of sunlight?

....

Which colors were brightest?

....

Did you find a pot of gold at the end of your rainbow?

....

Water Defies Gravity

Did the water stay in the glass?

....

How long did the water stay in the glass?

....

What is the best kind of card to use, thick or thin?

....

Did you get in trouble for soaking the carpets?

....

Testing Archimedes

How much water overflowed when you put the apple in the large jug?

....

What's the best way to put the apple in the jug, slowly or quickly?

....

What else did you put in the big jug?

....

What happens if you put your hand in the big jug?

....

Can you think of an experiment that would let you measure the displacement of your whole body?

....

Spinning Spiral

Did your spiral spin?

....

Where is the best place to hold your spiral if you want it to spin?

....

Which way did your spiral spin, clockwise or anticlockwise?

....

Did your cat like your spinning spiral, why is that?

....

Ice Magic

Did the ice cube stick to the thread?

....

How much salt should you use, a lot or a little?

....

Is it easier or harder with a bigger ice cube?

....

How amazed were your friends?

....

Snow Business

How long did you wait before looking under the cloths?

....

What happened to the snow under the dark cloth?

....

What happened to the snow under the light cloth?

....

What would happen if you waited longer?

....

What would happen if you used silver foil instead of cloth?

....

Does it matter how thick the cloth is?

....

Let's go to the Flicks

How many pages did you draw?

....

How long is your movie when you flick through it?

....

How many pages would you have to draw to make a movie an hour long?!

....

Who is starring in your next movie, Tom Cruise or a stick man?

....

Gas Gun

How far did the cork shoot out of the bottle?

....

How could you make it shoot further?

....

What is the most difficult part of this experiment?

....

How many windows did you break?!

....

OOOPS!

Moving Forward

Did your boat move?

....

How far did your boat move?

....

What substance made your boat move faster, washing-up liquid or oil?

....

What happens if you make the slot in the back of the boat bigger or smaller?

....

Could you sail your boat across the Atlantic Ocean?

....

QUICK, DIVE, IT'S A GIANT HAND

Take a Dive

Did your diver dive?

....

Did your diver go all the way to the bottom?

....

Why does the diver sink when you press your hand over the top of the jar?

....

Tower of Inertia

How high can you make your tower and still be able to knock out the bottom counter without it falling?

....

What happens if you use heavier or lighter counters?

....

How many times did your tower fall over before you got sick of this experiment?

....

GRRRRR!

Spud Balance

Did your potato balance?

....

Can you make the potato balance using only one fork?

....

What happens if you stick the forks into the potato nearer the top?

....

How long does the potato stay balanced?

....

Did you have chips for dinner?!

....

ANY MORE RESULTS, PICTURES OR IDEAS? PUT THEM HERE!

GREAT SCIENTISTS

GREAT SCIENTISTS

Ampère, André Marie (1775–1836)

French physicist who studied electricity and magnetism. How shocking! Gave his name to the ampere or amp, the unit of measurement for electrical current.

Archimedes (287–212 BC)

Greek mathematician and engineer. Particularly famous for running through the streets naked – nothing scientific about that!

BELL

Bell, Alexander Graham (1847–1922)

Scottish-born physicist and inventor. Invented the telephone – that's why his name always rings a bell.

Bernoulli, Daniel (1700–1782)

Swiss physicist and mathematician. He was the first person to use the letter "g" to represent the acceleration due to gravity. Gee!

ROBERT BOYLE

Boyle, Robert (1627–1691)
Irish-born physicist and chemist who is often regarded as the "father of modern chemistry." Hi, dad!

SCIENTISTS A to Z

Berzelius, Jöns (1779–1848)
Swedish chemist who discovered selenium (1818), silicon (1823), and thorium (1829), and made many other important contributions to chemistry. Obviously quite a brainbox.

Black, Joseph (1728–1799)
Scottish physicist who introduced the idea of latent heat after discovering that when ice is heated it melts with no rise in its temperature until it becomes water.

Bunsen, Robert Wilhelm (1811-1899)

German chemist after whom the Bunsen burner was named. Must have been quite a fiery character.

Carnot, Nicolas (1796-1832)

French physicist who did a lot of work on the theory of steam engines. It must have been frustrating for him as his work was ignored until 25 years after his death.

Cavendish, Henry (1731-1810)

English scientist who discovered hydrogen in 1766. He also discovered nitric acid and may have even liked eating acid drops.

Celsius, Anders (1701-1744)

Swedish astronomer who devised the Celsius (or Centigrade) temperature scale in 1742. Pretty hot stuff, eh?

Charles, Jacques Alexandre (1746-1823)

French physicist who built the first gas-filled balloon. That must have been a gas!

Copernicus, Nicolaus (1473-1543)

Polish astronomer who laid the foundations of modern astronomy. His findings that the planets orbited the sun were revolutionary.

COPERNICUS

Curie, Marie (1867–1934)

Polish-French chemist who discovered polonium and radium in 1898. She was awarded a Nobel Prize in 1903 and in 1911 became the first person to win a second.

Davy, Humphry (1778–1829)

British chemist who discovered sodium and potassium in 1807 and magnesium, calcium, barium, and strontium in the following year. He also invented a safety lamp for miners. Much of his work was developed further by his assistant Faraday, not to be confused with Friday who was Robinson Crusoe's companion.

Darwin, Charles (1809–1882)

British naturalist who claimed that man had sprung from apes. If you look at some examples of modern man you would come to the conclusion that many of them did not spring very far.

DARWIN

Edison, Thomas Alva (1847–1931)

Prolific American inventor. Invented the phonograph, the electric light bulb, and many other things. A real clever clogs.

Einstein, Albert (1879–1955)

German-born physicist famous for his theory of relativity – which has got nothing to do with his family.

EDISON

Faraday, Michael (1791–1867)

English physicist who produced the first electricity generator in 1831. A real dynamo.

EINSTEIN

Fermi, Enrico (1901–1954)

Italian physicist who built the first nuclear reactor. He also worked on the atomic bomb, proving that he was a really explosive character.

FARADAY

Foucault, Jean (1819–1868)

French physicist who made the first accurate measurement of the speed of light. It came to him in a flash.

Fresnel, Agustin Jean (1788–1827)

French physicist who invented the Fresnel lens used for car headlights. Bet he had a dazzling career.

Franklin, Benjamin (1706–1790)

American statesman and scientist who invented the lightning conductor after flying a kite in a thunderstorm. A real high flier.

Gabor, Dennis (1900–1979)

Hungarian-British physicist who invented holography in 1947. It was not until the invention of the laser in 1960 that holography became a feasible proposition.

FOUCAULT

Galilei, Galileo (1564–1642)
Italian scientist and astronomer. A really bright star of science.

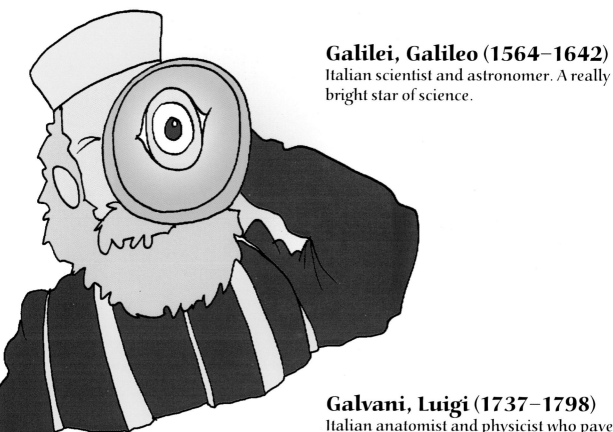

GALILEO

Galvani, Luigi (1737–1798)
Italian anatomist and physicist who paved the way for the development of the first electric cell. He did a lot of experiments with frogs' legs but never cooked them with enough garlic.

Gilbert, William (1544–1603)

Carried out many experiments with magnets and magnetism. Often regarded as an attractive person.

Hertz, Heinrich (1857–1894)

German physicist who discovered radio waves in 1888. Next time you drop a radio on your foot shout, "That Hertz."

Halley, Edmond (1656–1742)

British astronomer who accurately predicted the return in 1758 of the comet that now bears his name. He also published a map of the winds in 1686 – perhaps he'd eaten too many baked beans.

Hooke, Robert (1635–1703)

English physicist who discovered a law describing the elasticity of springs and formulated Hooke's Law (which has got nothing to do with fishing).

HOOKE

Joule, James Prescott (1818–1889)

British physicist who proved that a given amount of mechanical work produces a definite amount of heat. A unit of energy, the Joule, was named after him so when someone talks about jewels make sure you know what they are talking about.

Lavoisier, Antoine (1743–1794)

French chemist who got so involved he lost his head – he was guillotined.

OERSTED

Oersted, Hans (1777–1851)

Danish physicist who discovered electromagnetism.

Priestley, Joseph (1733–1804)

Conducted many experiments in electricity and was believed to have been the first person to isolate oxygen (although Karl Scheele actually beat him to it). Although Priestley his work was not connected with the church.

Leeuwenhoek, Anton van (1632–1723)

Dutch scientist who made his own single lens microscopes which enabled him to make detailed drawings of minuscule creatures. A pen also helped.

Lister, Joseph (1827–1912)

English surgeon and pioneer of antiseptic surgery. Not only did he cut out flesh, he also helped cut out disease.

Roentgen, William (1845–1923)

German physicist who discovered X-rays in 1895. He didn't know what they were so he used X, the symbol for something unknown in algebra.

Newton, Isaac (1642–1727)

English mathematician and physicist who formulated the theory of gravity. His mother, cooking lunch in the kitchen, was more interested in the law of gravy than the law of gravity (except when she dropped the gravy).

Rutherford, Ernest (1871–1937)

New Zealand-born British physicist who worked on radioactivity, which is not the same as activity heard on the radio.

Scheele, Karl (1742–1786)

Swedish chemist who actually discovered oxygen before Priestley but did not get it published first. A breath of fresh air in the scientific world.

Tesla, Nikola (1856–1943)

Croatian-American physicist who invented the induction motor in 1888.

Torricelli, Evangelista (1608–1647)

Italian physicist who is credited with the first man-made vacuum. He did not clean up with it but he did go on to invent the barometer.

Volta, Allessandro (1745–1827)

Italian physicist who invented the first electric battery. Unfortunately he did not have a TV remote control to put it in. Gave his name to the volt, the unit of measurement of electromotive force.

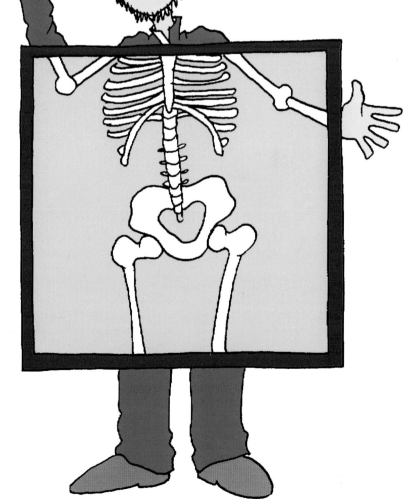

ROENTGEN